FERNS AND FERN ALLIES
of New Zealand

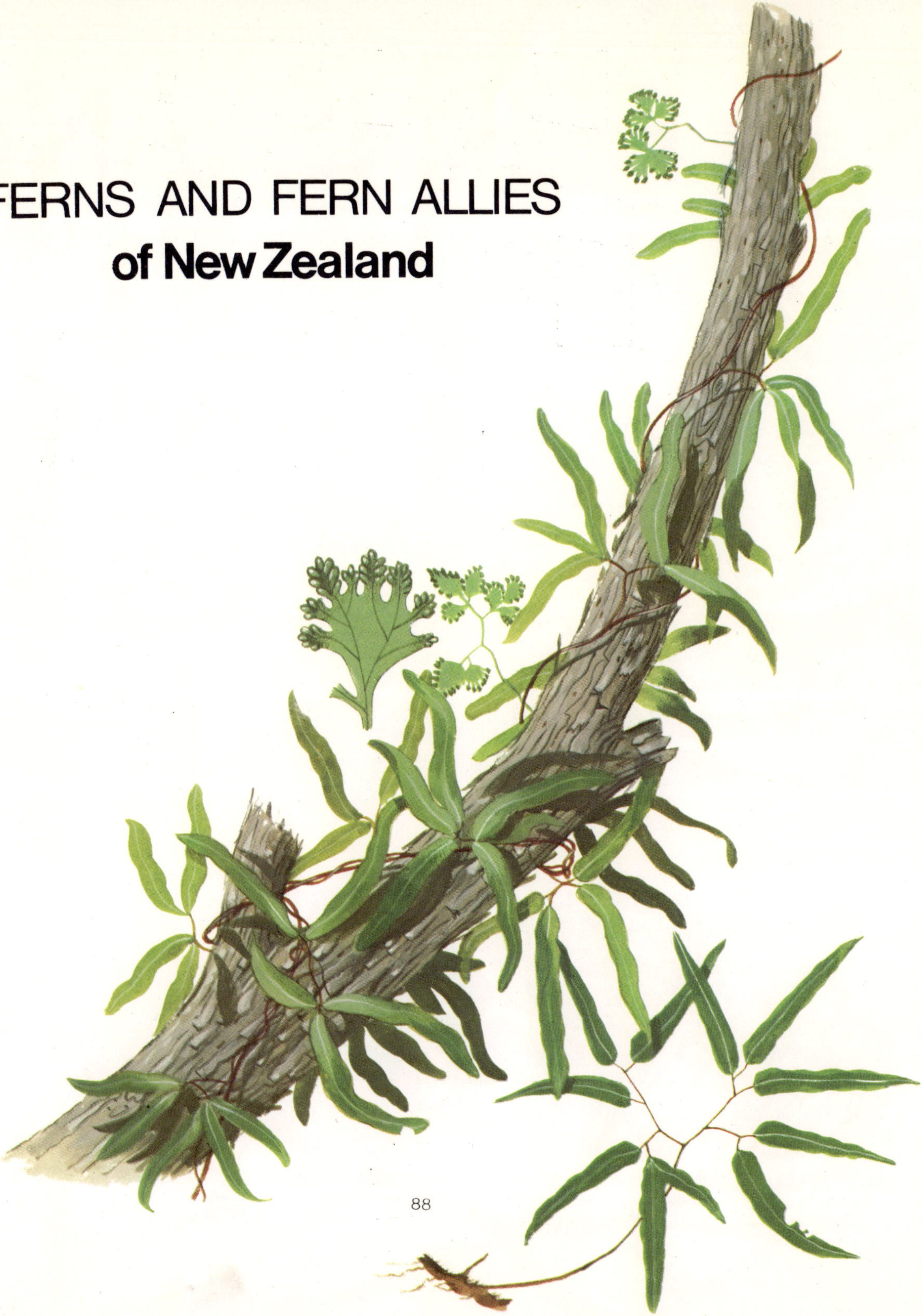

FERNS AND

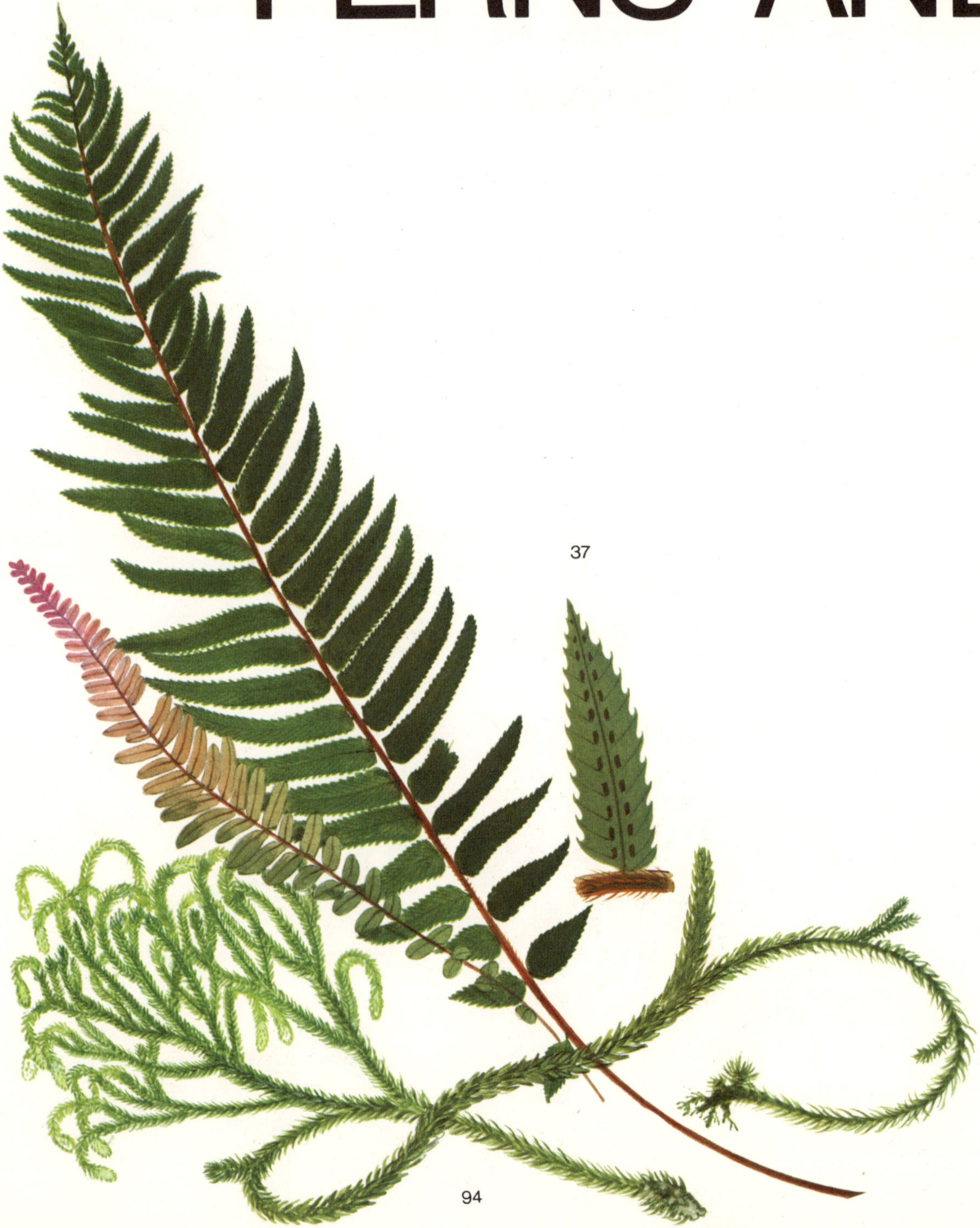

FERN ALLIES
of New Zealand

Designed and illustrated by
ERIC HEATH

Text by R.J. CHINNOCK

A.H. & A.W. REED
Wellington · Sydney · London

First published 1974

A.H. & A.W. REED LTD
182 Wakefield Street, Wellington
51 Whiting Street, Artarmon, NSW 2064
11 Southampton Row, London WC1B 5HA
also
29 Dacre Street, Auckland
165 Cashel Street, Christchurch

ISBN: 0 589 00865 X
Library of Congress Card Catalogue No: 74-76512

Printed by Kyodo Printing Co. Ltd. Tokyo

ACKNOWLEDGEMENT

The authors wish to acknowledge the help received from Mr B.V. Sneddon, Botany Department, Victoria University of Wellington.

CONTENTS

INTRODUCTION

WE ARE FORTUNATE in New Zealand in having a mild climate with ample rainfall which is well suited to the growth of Pteridophytes, and in most parts of the country they abound in large numbers. The forest, with its protective mantle of trees, and broad range of habitats, is their most popular haunt and a walk along any forest track or stream will provide you with many interesting and varied species to study.

Although one usually associates Pteridophytes with moist, shaded situations there are some species which prefer the solitude of exposed coastal cliffs, dry sunny hill-slopes, swamps and even open rocky mountain places. Even in one of the most arid parts of the country like Central Otago a few hardy rock-ferns including *Cheilanthes sieberi* (8) are found.

There are 171 Pteridophytes currently recognised in New Zealand consisting of 152 ferns and 19 fern-allies; of these we have described and illustrated 88 ferns and varieties and 18 fern-allies. In most books on New Zealand ferns the fern-allies have been ignored so in this book we include illustrations and descriptions of all but one species. Most of the common Pteridophytes are widespread throughout the country, but a few like *Lygodium articulatum* (88), *Lycopodium deuterodensum* (95) and *Tmesipteris lanceolata* (106) although restricted to certain parts of the country are common within those areas except near the southern limit of their range.

For each species a description of the more important features used in its identification together with notes on the habitat and its distribution are given. All measurements are in millimetres (mm) or centimetres (cm) or metres (m). Magnifications given are the number of times larger or smaller than the natural size (X1), i.e. X5 – five times the natural size; X0.2 – one fifth natural size. A 10X hand lens is recommended for the examining of small structures such as the sorus, scales and hairs.

FERN CLASSIFICATION

FERNS, like all plants, have a scientific binomal name. The first name is that of the genus, the group to which the plant belongs and the second, the species name, identifies the plant from other members of the genus. All plants of the same species have the same general appearance. Their frond form and the sorus type and position is the same but they may vary slightly in frond size and the degree of hairiness depending on where they are growing. The species of a genus may differ considerably in their frond shape, pinnae form and their mode of growth but they are all grouped together because they have a number of basic features in common which are characteristic of that genus.These features are usually related to the soral structure. Thus all the species of *Asplenium* have a characteristic elongated sorus which is oblique to the mid-vein and covered by an indusium which opens towards the vein.

Unfortunately, as our knowledge of genera and species increases it often becomes necessary for botanists to change names. These name changes sometimes confuse the amateur naturalist who merely wants a name for a species and he objects when a name he has known for years is altered. Botanists follow a code of ethics, "The International Code of Botanical Nomenclature", the prime aim of which is to provide a stable classification. A name change can occur for many reasons. If, for example, a plant was found to have been given an earlier validly published name than the one currently used then the earlier name is the valid one although the latter name may have been in use for years. Likewise, if a species is transferred to another genus in which there is already a species with the same specific name then the transferred species must be given another specific name.

Colloquial and vernacular names have only been included if they are well known and in common usage. A large number of New Zealand ferns do not have common names and no attempt has been made to coin any. Common names are often misleading because a species may have a number of common names or it may share a common name with other species which may or may not be closely related. Such names as "filmy-fern" and "tree-fern" do not mean much when you consider that we have 28 filmy-ferns and 8 tree-ferns. The scientific name on the other hand, although it may be a mouthful, does convey the affinities of of the species and very often the specific name denotes a feature of the plant.

FERN IDENTIFICATION

BEFORE we can hope to identify Pteridophytes there are a number of basic terms that we must understand. The ferns are discussed below and the fern-allies separately in their own sections. Composite Figure 1 illustrates the more important features of the pteridophytes, and a glossary of botanical terms is to be found at the back of the book.

The Fern Stem is called the rhizome and its mode of growth is very important for distinguishing species. It may be slow-creeping with crowded fronds (Fig 1A), long-creeping with scattered fronds (Fig 1B), short and condensed with the fronds clustered at its tip (Fig 1C), or erect and forming a trunk with the fronds forming a crown. Stolons may be produced by the main stem and these creep along the ground and eventually root at the tip and produce a new plant.

The Fern Leaf is generally termed the frond. It consists of a stipe (stalk) and lamina (blade). The midrib of the lamina is called the rachis and this may differ markedly from the stipe in colour, shape and covering of hairs and scales. The frond may be simple and undivided (Fig 1C) or divided into a number of basic divisions called pinnae (Fig 1A). The pinnae may be arranged alternately or in opposite pairs along the rachis. If the pinnae are undivided as for instance in *Pellaea rotundifolia* (6) the frond is termed pinnate. More commonly, however, the pinnae are further divided once or twice into secondary and tertiary pinnae (Fig 1A). A frond in which the pinnae are divided twice into distinct stalked segments is termed bipinnate, and thrice, tripinnate. The ultimate stalked segment of a pinna is termed the pinnule. The pinnule may be variously lobed or toothed.

The Sporangia, their arrangement, position and associated structures are essential for an accurate identification. If we look at the underside of a mature frond we will see spots or lines at various positions. These spots or lines are the sori and if we take a closer look at a sorus with a 10× hand lens we will see that it is composed of a number of small globular bodies called sporangia. The sporangium is the sac in which the spores develop and mature. The structure of the sporangium varies considerably and it is used by botanists to break down the ferns into their major divisions. Primitive ferns such as *Botrychium* have large thick sporangia with many hundreds of spores while advanced ferns such as *Anarthropteris lanceolata* have small thin sporangia with specialised opening mechanisms and no more than 64 spores.

The sorus may be exposed on the surface of the frond or it may be variously covered by an indusium. The indusium varies considerably in shape and form and the term as used in this book has been broadened to include reflexed pinnule margins of the type found in *Adiantum* and *Pteris.*

Hair and scale coverings of the rhizome and frond are often of importance for distinguishing species and genera. Hairs vary in shape and length and may be simple, consisting or one or a single row of cells (Fig 1E), or complex like the glandular (Fig 1E and stellate (see 61a) types. Scales also vary considerably and their colour, shape and cell arrangement is often useful. In the tree-fern genus *Cyathea* for example *C. medullaris* is easily distinguished from our other species by its toothed scale which has all the cells orientated in the same direction. The other species have scales with entire margins and the cells near the margins orientated in different directions. The other tree-fern genus *Dicksonia* is easily distinguished from *Cyathea* because its indumentum consists of hairs only.

Variation in the frond size, its harshness and degree or hairiness is largely dependent upon where the plant is growing while the sorus and its associated structures and the general structure of the frond are rarely affected by the environment. In some species like *Microsorium diversifolium* (12), *Asplenium flaccidum* (19), and *Hymenophyllum multifidum* (56) the frond size variation is very large and it is very difficult at first to accept that such extremes as found in these species all belong to that species. On coastal cliffs, for example, the mature fronds of *Microsorium diversifolium* may be as small as 5-6 cm while in the forest the fronds are eight to ten times that size. The size of the fern frond therefore should not be relied upon too greatly for identification. Instead the soral features and the general composition of the frond should be used. In general, species that are found under a wide range of habitats are the most variable.

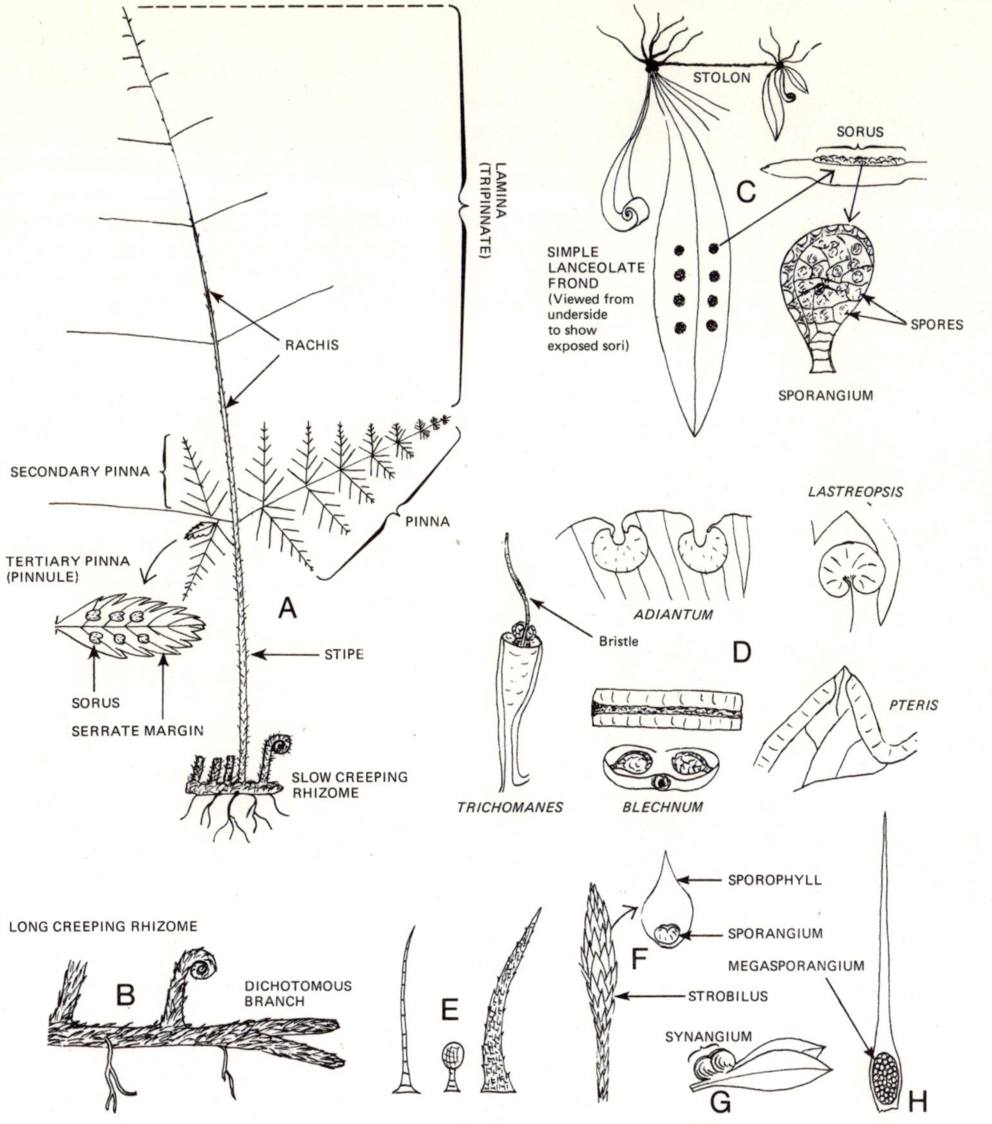

Figure 1. Important features of the Fern and the Fern-allies.

A. A tripinnate frond of *Lastreopsis hispida* (diagrammatic) showing various important features. B. A long creeping rhizome of *Gleichenia.* C. Plant of *Anarthropteris lanceolata* illustrating its simple lanceolate fronds, stoloniferous habit, the sorous and sporangium. D. Sorus coverings: *Trichomanes*, tubular indusium; *Blechnum*, marginal indusium on modified pinna; *Lastreopsis*, kidney-shaped indusium; *Pteris* and *Adiantum*, modified reflexed margin of the pinnule. E. Scale and hairs: a simple hair (left), a glandular hair (centre) and a scale (right). F. Strobilus of *Lycopodium* with one sporophyll removed and enlarged to show the kidney-shaped sporangium near its base. G. Lateral branch of *Tmesipteris lanceolata* illustrating the synangium, which consists of two fused sporangia, and the associated leafy appendages. H. A megasporophyll of *Isoetes* with the megasporangium at its base.

THE FERN LIFE CYCLE

THE COMMON FERN that we are all familiar with comprises only the asexual phase of the life cycle of the plant so we should look briefly at the life cycle as a whole (depicted in Fig 2). The life cycle of the fern-allies is essentially the same.

The typical fern plant, the sporophyte, produces asexual spores on the undersurface of its fronds. After the spores are released from the sporangium they fall to the ground or they may be blown about for a while before they settle. If the ground is suitable the spores will eventually germinate and develop, not into a plant like the "parent", but into a small heart-shaped body, the gametophyte (or prothallus), which rarely exceeds 8 mm in breadth. The gametophyte is the sexual phase in the life cycle, and even the mighty Mamaku (*Cyathea medullaris*), one of the tallest ferns in the world, is humbled by having a gametophyte no larger than those of many of our smallest ferns. The gametophyte is very thin and delicate and to survive it must have a continual supply of moisture so it is usually found growing on damp banks, on rocks near streams and on the damp bases of trees and tree-ferns.

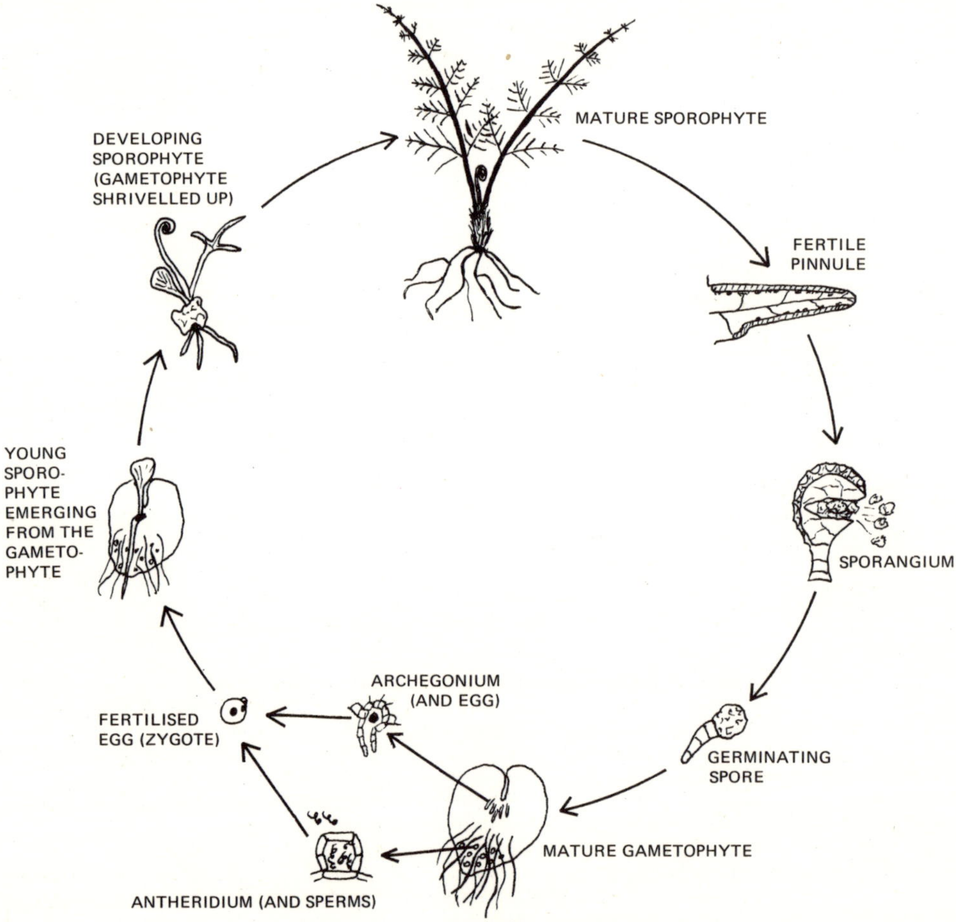

Figure 2. The Fern Life Cycle (Diagrammatic)

As the gametophyte develops, sex organs form on its undersurface. The male ones, the antheridia, develop in the lower part of the body amongst the rhizoids (water absorbing hair-like structures which also act as an anchor) while the female organs, the archegonia, develop around the notch in the upper part of the body. When the antheridium is mature its cap cell ruptures and the sperms are released to swim about in the water. The archegonium, when mature, is flask-like and it has a single egg at its base. Sperms are attracted to the archegonium by a fluid which it exudes and eventually a sperm finds its way into the archegonium, fuses with the egg and a zygote (fertilised egg) is formed. When this happens all the other egg cells of the gametophyte cease to function. As soon as the zygote divides it is called an embryo. The embryo gradually increases in size and takes its nourishment from the gametophyte body. Eventually a small simple frond and a root of the developing sporophyte emerge from the gametophyte. The young sporophyte is now independent as it can take its own nourishment from the soil and produce food through photosynthesis, and the gametophyte, whose purpose is now fulfilled, gradually shrivels up and dies. The time taken for a young sporophyte, which has just emerged from the gametophyte, to develop into a mature spore-producing plant varies considerably from species to species. Mature fronds of Annual Fern (5) are produced in a couple of months while many tree-ferns take in excess of twelve years to produce spore-bearing fronds.

MAIDENHAIR FERNS
and others

1. Adiantum cunninghamii. "Common Maidenhair Fern". Rhizome creeping, scaly. Stipe black, shining, glabrous except near the base. Lamina of 3 to 8 alternate pinnae; pinnules (1a) oblong, dark green above, glaucous below. Sori in the notches between the lobes (1b), indusium glabrous. (**1.** x0.3; **1a.** x1.2; **1b.** x6)

Abundant throughout the country in drier situations in lowland forest and bush remnants. Dwarfed forms with few pinnae are often encountered on rocky banks along stream courses in open places.

2. Adiantum aethiopicum. "True Maidenhair Fern". Rhizome thin, wiry, long creeping and producing tufts of fronds at intervals. Fronds light green 15-50 cm long. Fertile fronds annually produced during spring and summer, dying off in autumn. Pinnules rounded, sori few, large (2a). (**2.** x0.3; **2a.** x6)

Common in drier open places in the northern North Island from near North Cape to Hamilton but most frequent north of Auckland. South of Hamilton it is very rare and local and is known only from a few localities. It is recorded from Nelson in the South Island.

3. Adiantum diaphanum. "Small Maidenhair Fern". Fronds tufted together, small, rarely more than 20 cm long. Lamina very dark green, pinnate, or with one or two basal pinnae. Pinnules oblong (3a), sori small, indusium covered with bristles (3b). (**3.** x0.8; **3a.** x1.5; **3b.** x8)

Found throughout the country but often local. Usually growing on moist shaded banks and rocks near streams.

4. Adiantum hispidulum. "Rosy Maidenhair". Rhizome short, scaly. Fronds tufted; stipe black, rough, shiny. Lamina fan-shaped, usually hairy. Sori numerous, indusium covered with bristles. Young developing fronds, in open places, are usually rose-red. (**4.** x0.3)

Common from North Cape to Hamilton on banks and in open rocky places.

5. Anogramma leptophylla. "Annual Fern". Fronds few, tufted together, rarely more than 8 cm long, glabrous. Sori unprotected, forming bands along the veins of the pinnae (5a). (**5.** x1; **5a.** x3)

Known in scattered localities from Auckland to Dunedin. Often overlooked because of its small size and annual habit. New plants start developing from April onwards, mature in late winter and start dying off as the banks dry out in late Spring. I have always found this species growing on south- or east-facing vertical banks or rocks in open places, usually associated with mosses and liverworts.

6. Pellaea rotundifolia. Rhizome slow creeping, fronds tufted together. Stipe and rachis covered in brown scales. Pinnules rounded or elongated. Sori forming a band along the margin of the pinnule but not reaching the tip or base. (**6.** x0.5)

Quite common in drier forest situations throughout the country. Occasionally encountered in open rocky places.

7. Cheilanthes distans. Fronds tufted, 10-20 cm long, 2-3 cm wide. Stipe, rachis and undersides of pinnae covered with buff-coloured to brown scales. Upper surface of pinnae deep green, scaly or glabrous. Sori forming bands along the margins of the pinnule segments. (**7.** x0.5)

A sun-loving fern that is common on coastal cliffs and inland rocky places from Northland to Canterbury.

8. Cheilanthes sieberi. Fronds tufted, 8-20 cm long and 2-3 cm wide, occasionally wider in shaded forms, dark green, glabrous. Sori as for *C. distans.* (**8.** x0.5)

Found throughout the country in similar situations to *C. distans* (**7**) and often associated with it.

9. Azolla rubra. "Red Azolla". Plants 1-2.5 cm long, green to purplish-red, reproducing largely by fragmentation and often covering large expanses of water. (**9.** x1)

Abundant everywhere and often forming large "red carpets" on ponds, lakes and slow moving waters. Occasionally found on soil near water seeps.

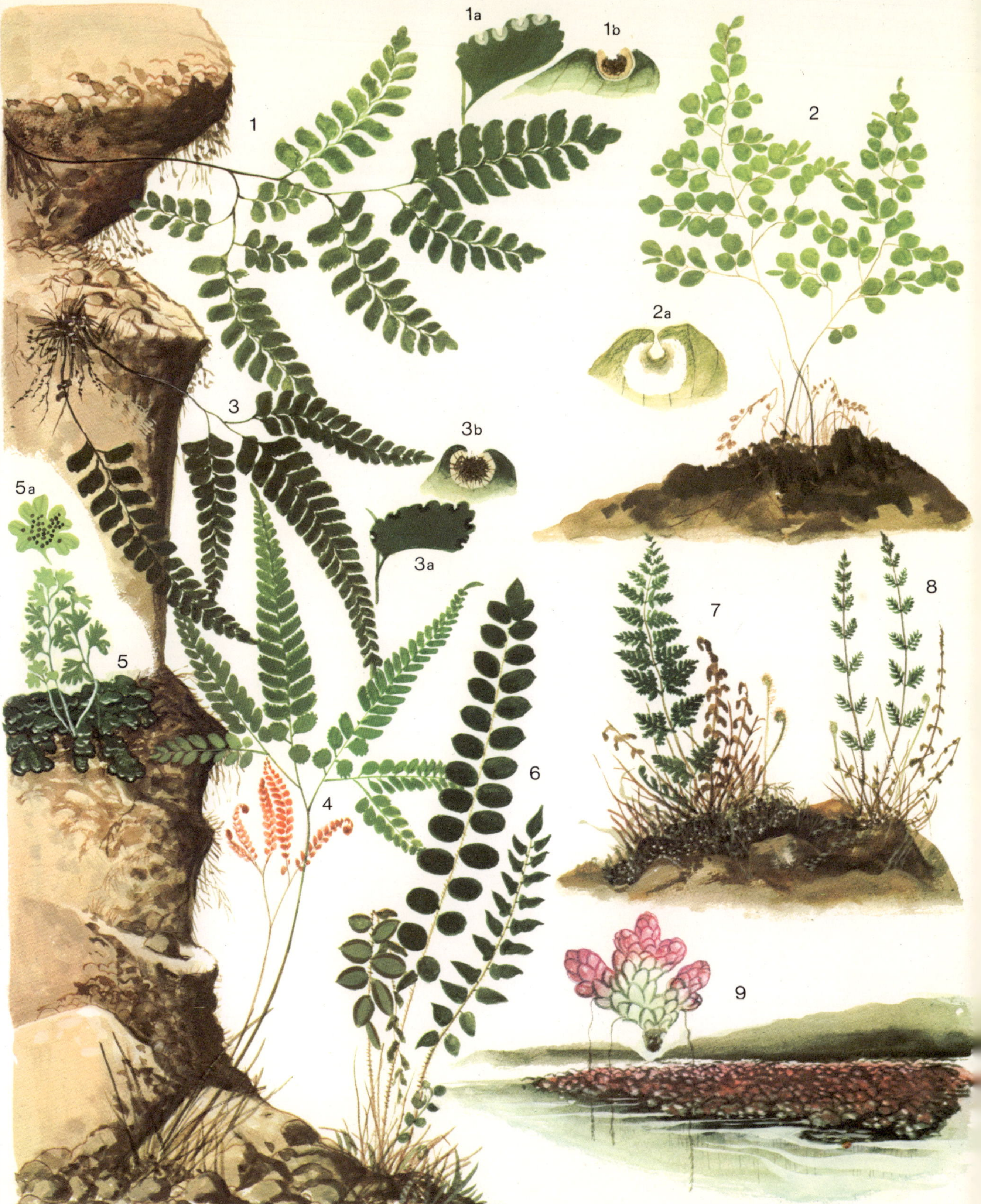
1a
1b
1
2
2a
3
3b
5a
3a
8
7
5
6
4
9

FRAGRANT FERN
and others

10. Pyrrosia serpens. Rhizome long-creeping, much branched, covered with brown scales. Fertile fronds narrow, tongue-like, 4-12 cm long, 1-2 cm wide. Sterile fronds smaller, often rounded. Lamina thick, fleshy, green to glossy-green above, densely clad in buff-coloured matted stellate hairs below. Sori in two rows. (**10.** x0.5)

A hardy sun-loving fern, catholic in its taste, and abundant throughout the country in the forest (usually in the tops of trees), under scrub, on rocks, coastal cliffs, rock walls and on exotic trees.

11. Microsorium scandens. "Fragrant Fern". Rhizome long-creeping, slender, 3-4 mm diameter, covered with brown scales. Fronds simple or pinnatifid 10-35 cm long, membranous, dark green. Sori numerous, 1-3 mm diameter, naked. (**11.** x0.5)

A very common forest fern in most parts of the country except the eastern side of the South Island. Absent from Stewart Island. Common on trees, banks and the forest floor.

12. Microsorium diversifolium. Rhizome long-creeping, thick, fleshy, 6-10 mm diameter, partially covered with brown scales. Fronds simple, unevenly pinnatifid or pinnatifid, 10-60 cm long, bright green, leathery. Sori numerous, 2-4 mm diameter, sunken and appearing as spots raised on the upper surface. (**12.** x0.5)

An abundant fern throughout the country at all altitudes. It has no particular tastes and may be found on dry coastal cliffs, open rocky places, in the forest as an epiphyte or ground fern, and also on walls and exotic trees.

13. Ctenopteris heterophylla. (*Grammitis heterophylla*). Fronds tufted together, dark green, leathery, pinnate or pinnatifid in the forest forms 8-25 cm long; or simple with irregular toothed margins in exposed rock growing forms, which rarely exceed 6 cm in length, and young plants. Sori oblong, oblique, 2-3 mm long. (**13.** x0.8)

A common epiphyte in forest throughout the country at all altitudes. Occasionally found growing on rocks and banks in open situations.

14. Grammitis billardieri. Rhizome slow creeping, live fronds tufted together. Fronds linear-lanceolate, 2-10 cm long, glabrous. Sori oblong, oblique to the midrib. (**14.** x0.8)

A small forest fern found in all parts of the country particularly in Beech *(Nothofagus)* Forests. Usually found at the bases of trees and on banks. Another closely related species, *G. ciliata*, is distinguished from *G. billardieri* by the presence of long brown or white hairs on the stipe and underside of the frond.

15. Anarthropteris lanceolata. Rhizome condensed, scaly, emitting long woolly rootlets which produce tufts of new fronds at intervals. Fronds lanceolate, firm, erect or hanging, 5-25 cm long. Sori very large, sunken, visible as raised spots on upper surface. (**15.** x0.5)

Common as an epiphyte on tree trunks, on rocks or on banks in lowland forest throughout the North Island and the north and west of the South Island as far south as Greymouth. It is particularly abundant in Northland.

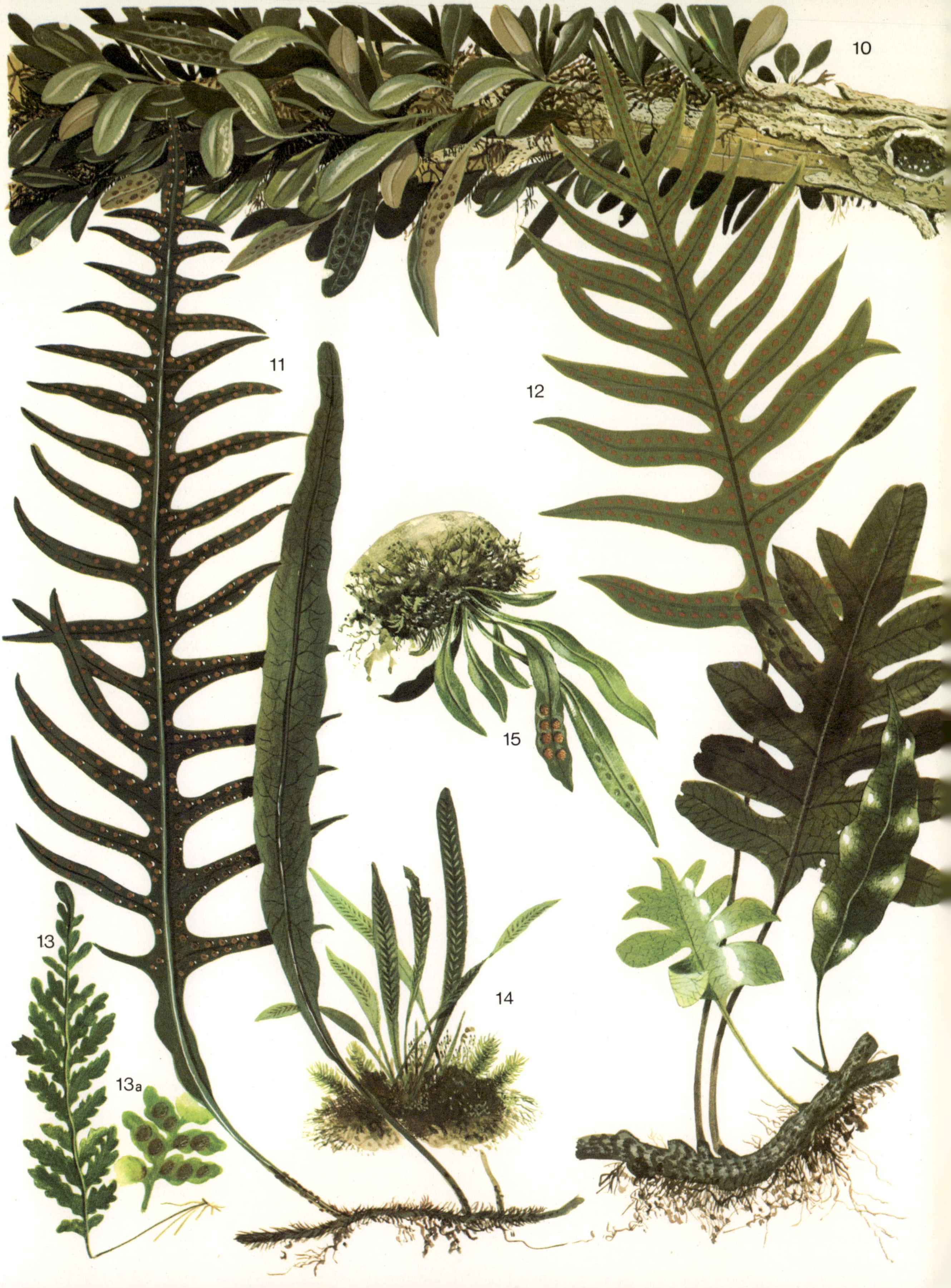
10
11
12
15
13
14
13a

SPLEENWORTS

16. Asplenium lucidum. "Shining Spleenwort". Rhizome slow creeping, covered with long brown scales. Fronds pinnate, 20-100 cm long, erect or drooping. Pinnae broad lanceolate, bright glossy green above, dull green below, firm and leathery, margins finely serrate. Sori numerous, linear, 1-2 cm long, oblique to midrib (16a), indusium flap-like, delicate, obscure in old sori. (**16.** ×0.3; **16a.** ×0.8)

Abundant throughout the country in lowland to montane forests as an epiphyte or ground fern. Easily identified by its bright shining pinnae.

17. Asplenium bulbiferum. "Hen and Chicken Fern". Rhizome short, usually erect. Fronds 20-150 cm long, pale to dark green, bipinnate to tripinnate. Pinnules variable in degree of dissection. Bulbils abundant, scattered or few on the upper surface of the lamina, often absent in young plants. Sori broad oblong, near the margins. (**17.** ×0.3)

A very common forest fern particularly along water courses where it attains its greatest size. This species hybridises readily with a number of Spleenworts particularly with *A. flaccidum* (19) and evidence of this is seen in the occasional A. *"flaccidum"* plant sporting bulbils.

18. Asplenium flabellifolium. "Necklace Fern". Rhizome short, fronds few prostrate or sub-erect, very fragile. Rachis slender producing a long naked tip which roots and produces a new plant. Pinnae fan-shaped. Sori 2-6 per pinna, 3-4 mm long, oblique to midrib. (**18.** ×1; **18a.** ×2)

Usually found in rocky open places or on banks in open coastal forest from Kaitaia southwards. Exposed forms are often small with yellowish-green pinnae.

19. Asplenium flaccidum. "Hanging Spleenwort". (a) Forest form illustrated: Fronds 15-100 cm or more long, pinnate, flaccid, pendulous, thick and leathery. Pinnae entire, fluted, shallow or deeply lobed. (b) Coastal form: Fronds 2-40 cm long, firm, spreading or erect, bipinnate, often very fleshy. (**19.** ×0.5)

An abundant fern throughout the country in many situations from coastal cliffs to alpine rocky places. Our most variable New Zealand fern; poorly understood, and probably consisting of two to three species or subspecies.

20. Asplenium falcatum. Rhizome short creeping. Fronds 20-100 cm long, stipe and rachis scaly, pinnae lanceolate, margins incised, serrate, often deeply so. Sori numerous, oblique to midrib (20a). (**20.**, **20a.** ×0.8)

Common in forests throughout the country as an epiphyte on trees, or on rocks, fallen logs or the forest floor. An easily identified species because of its distinctive pinnae.

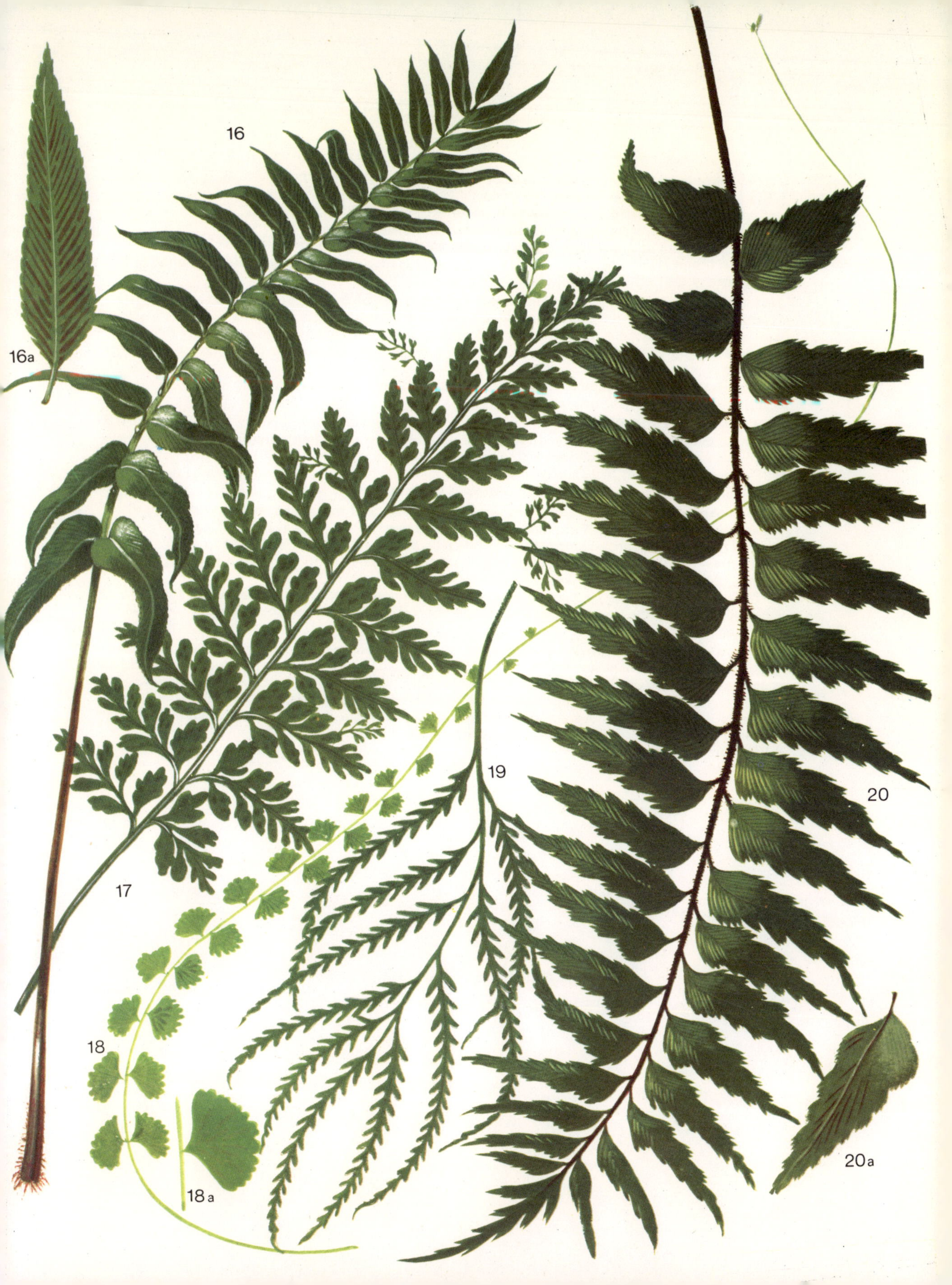
16
16a
17
18
18a
19
20
20a

SHIELD FERNS
and others

21. Polystichum vestitum. "Prickly Shield Fern". Rhizome erect, often forming a short trunk up to 1.5 m high. Fronds 30-100 cm long, dark green, bipinnate. Lamina very harsh, pinnules prickly. Stipe and rachis densely covered with brown to black-brown scales with lighter margins. Sori 4-8 per pinnule (21a), indusium thin, membranous, centre black. (**21.** ×0.5, enlargement ×5)

A common fern in cold wetter areas, becoming increasingly abundant southwards. In the North Island it is found mainly in mountain regions. This species often covers cleared hillsides and other disturbed sites to the exclusion of other plants.

22. Lastreopsis glabella. Rhizome short, fronds tufted. Stipe glabrous, lamina glabrous on upper side except for rachis and pinnae midribs which are covered with soft reddish brown hairs. Lower surface of lamina possessing scattered appressed hairs. Sori small, at first covered by a pinkish kidney shaped indusium but this is gradually pushed to one side by the developing sporangia (22a). (**22.** ×0.3; enlargement ×5)

Widespread throughout the country in lowland forest. This species prefers drier situations but is occasionally found growing on stream banks.

23. Lastreopsis hispida. Rhizome slow creeping, fronds variable in size, 30-100 cm long. Rhizome, stipe and rachis densely clad with dark reddish-brown to brown bristle-like hairs. Midribs of pinnae hairy. Sori 4-8 per pinnule (23a), indusium kidney-shaped. (**23.** ×0.5; enlargement ×6)

Common throughout the country in lowland to montane forests. Usually encountered on the forest floor but occasionally on the lower trunks of trees and tree-ferns.

24. Rumohra adiantiformis. Rhizome long creeping, densely clad with orange-brown scales. Fronds variable in size, usually 30-100 cm long. Stipe thick, grooved above, scaly at first but glabrous with age. Lamina firm, thick, glabrous, green to yellowish-green above, pale green below. Sori large, numerous, in two rows near the midrib. Indusium shield-like, falling off at maturity. Sporangia jet black when mature. (**24.** ×0.7; enlargement ×5)

Usually found climbing on trees but occasionally encountered on fallen logs or on the forest floor. Easily identified by the long creeping rhizome with its orange brown scales and the thick firm fronds with their distinctive sori.

25. Polystichum richardii. Rhizome short, erect, clad in dark brown scales. Fronds 20-50 cm long in coastal forms but over 100 cm long in some forest forms. Stipe covered with dark brown scales and scurfy hairs, but the scales often deciduous. Lamina dark bluish-green firm and often fleshy (in coastal forms), glabrous above but covered with scurfy hairs below to varying degrees. Sori large, numerous, covered with a shield-like indusium with a prominent black centre. Sporangia black at maturity extending beyond indusium (25a). (**25.** ×0.5; enlargement ×5)

Widespread throughout the North and South Island and most frequent near the coast on cliffs and under scrub.

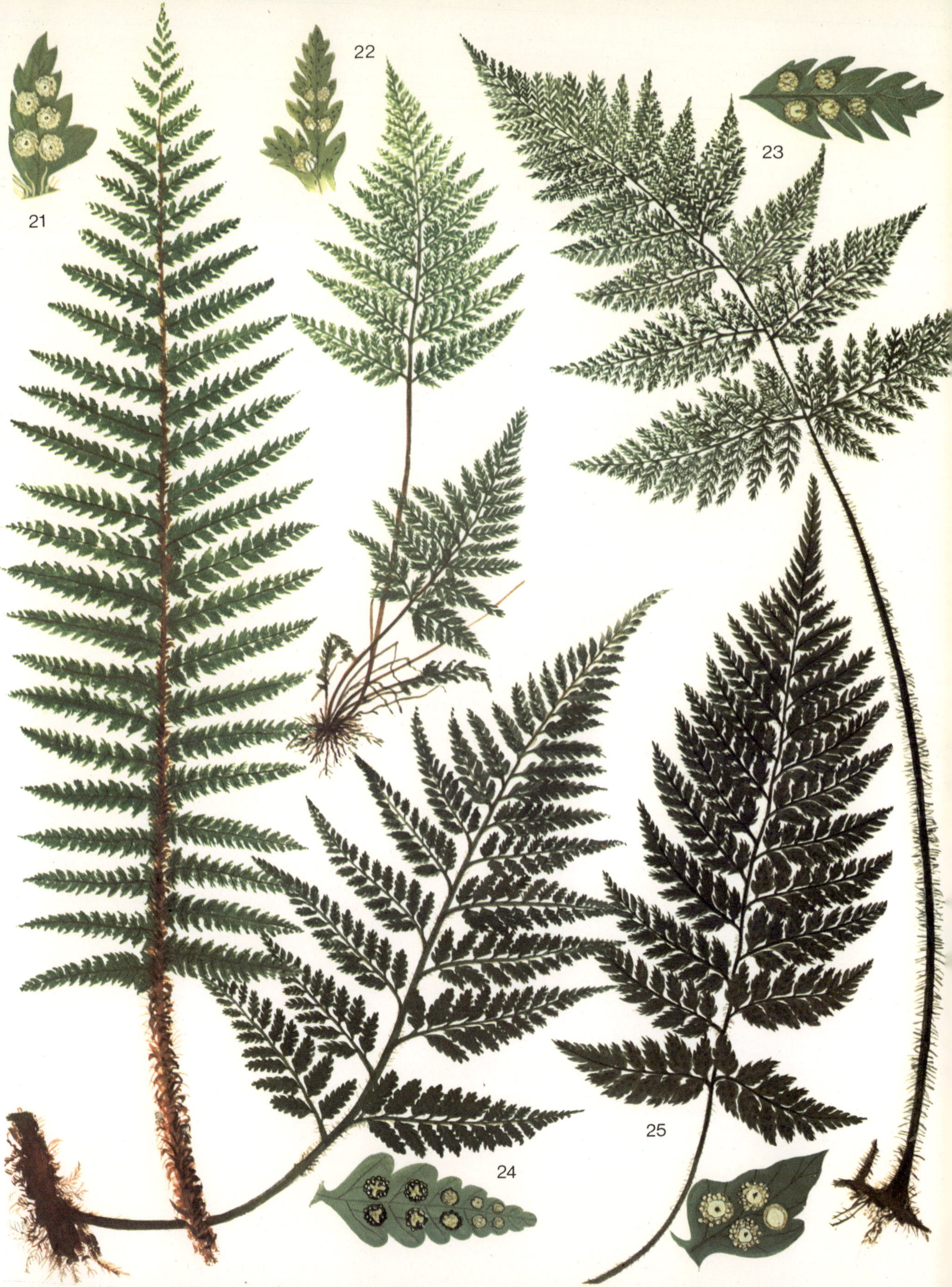

21
22
23
24
25

BLECHNUM FERNS

The Blechnum genus has distinct sterile and fertile fronds. Fertile fronds consist of modified narrow brown pinnae, the undersurface of which is covered with fused sori that form a continuous band either side of the midrib. The indusium projects out from the margin, extends over the sporangia, and almost meets the indusium from the other margin near the midrib. Sterile fronds only are illustrated for *B. vulcanicum* (28), *B. procerum* (33) and *B. fraseri* (36).

26. Blechnum filiforme. Rhizome long creeping. Ground fronds sterile, small, 6-15 cm long. Mature fronds on climbing rhizomes 20-50 cm long, the fertile fronds with very long thin drooping pinnae. (**26.** x0.2)

Abundant in lowland forests throughout the North Island and in the extreme north of the South Island. When rhizomes start to climb trees the fronds gradually increase in size. Fertile fronds are only developed on climbing rhizomes.

27. Blechnum penna-marina. Rhizome creeping. Fronds 10-30 cm long, 1-1.5 cm wide, often reddish when young. Fertile fronds erect, pinnae curved. (**27.** x0.5)

Widespread throughout the country particularly in mountain regions in open grasslands or on the margins of forest.

28. Blechnum vulcanicum. Rhizome creeping, fronds 20-70 cm long, lowermost pinnae pair usually strongly curved. Fertile fronds usually longer, similar in shape to sterile ones. Young fronds often bright red. (**28.** x0.5)

Found from Northland southwards particularly on banks. In the North Island it is more frequent in mountain regions. North of Auckland it is known only from near Kaeo.

29. Blechnum nigrum. Rhizome short, erect. Sterile fronds small, tufted, blackish-green with a very distinct large terminal pinna. (**29.** x0.5)

This small fern is found throughout the country but it is often local. It is usually found in gloomy places near streams or on damp banks.

30. Blechnum chambersii. (*Blechnum lanceolatum*.) Fronds tufted, lanceolate, 15-60 cm long. Pinnae 2-5 cm long gradually tapering. Young fronds often bright red. Fertile pinnae erect. (**30.** x0.2)

A common forest fern usually encountered along the banks of streams.

31. Blechnum membranaceum. Fronds small, dark green, pinnae 1-2 cm long, oblong to sub-orbicular. Young fronds often reddish-brown to red. (**31.** x0.2)

A small forest fern usually found along stream banks and often associated with *B. chambersii* with which it freely hybridises.

32. Blechnum fluviatile. Fronds tufted together, sterile ones spreading, deep green to brownish-green, pinnae rounded. Fertile fronds erect, pinnae projecting upwards. (**32.** x0.5)

Common throughout the country from Rotorua southwards and usually found near streams. More common in cooler mountain areas in the North Island.

33. Blechnum procerum. Rhizome creeping, fronds 15-60 cm long. Pinnae usually brownish-green to brown, thick, brittle. Terminal pinna longer than lateral ones, the lowermost pinnae pair shorter, curved. (**33.** x0.5)

Common along forest margins, in open forest, or on banks throughout the country.

34. Blechnum discolor. "Crown Fern". Fronds tufted, forming a crown. Sterile fronds bright green above, whitish below. Fertile pinnae narrow, the lowermost ones with broad sterile bases. (**34.** x0.3)

Common in forest areas throughout the country where it often covers extensive areas. More common in mountain areas in the North Island.

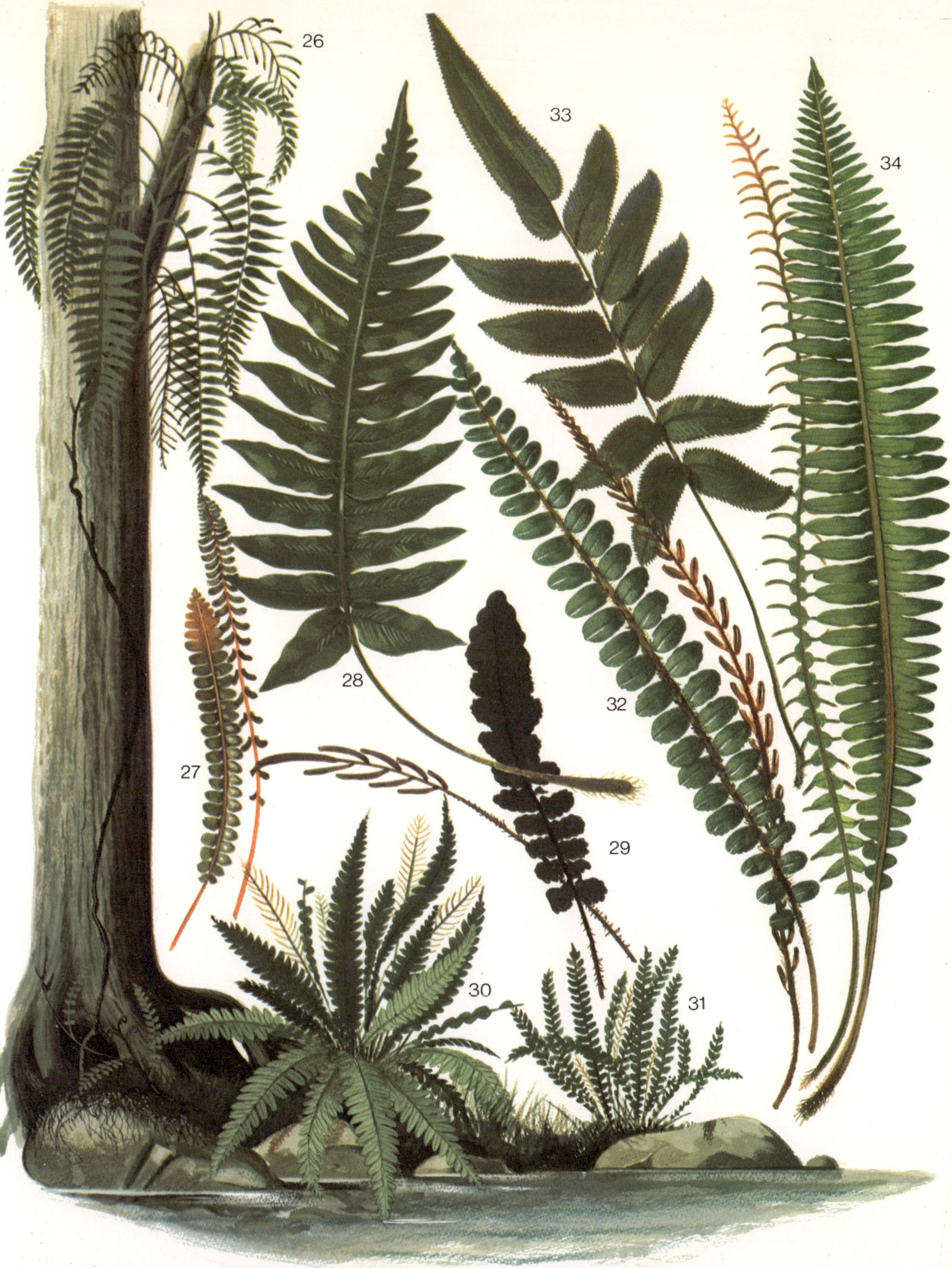
26
33
34
28
32
27
29
30
31

BLECHNUM FERNS
and others

35. Blechnum capense. "Kiokio". Rhizome creeping, covered with brown scales. Fronds lanceolate 20-60 cm long. Stipe scaly, particularly near the base. Pinnae lanceolate, margins finely serrate and often undulate. Lowermost pinnae small, rounded, almost clasping rachis. In open places young developing fronds are bright red. (**35.** x0.3)

In New Zealand this species is very complex and consists of four distinct taxa. The one illustrated is common on road cuttings and cliffs throughout the country.

36. Blechnum fraseri. Rhizome erect, forming small trunks up to 1.2 m high, 2-3 cm diameter. Propagates vegetatively by stolons. Fronds clustered together, forming a crown. Sterile and fertile fronds similar, although segments of fertile fronds narrower, bipinnate, lamina dark bluish-green above. (**36.** x0.4)

A common forest fern from near North Cape to south of Hamilton. It is absent from the central and southern North Island but occurs at scattered localities in the north west of the South Island.

37. Doodia media subspecies **australis.** Rhizome ascending, covered in scales. Fronds tufted, often bright red or purplish-red at first, becoming dark green and very harsh when old. Sori in a single or double row each side of the midrib. (**37.** x1; enlargement x4. Appears on title page).

A very common fern on dry open banks in the northern North Island. South of Rotorua it is known only from a few localities but extends southwards to Nelson and Marlborough.

38. Pteris macilenta. Rhizome short, erect. Fronds tufted, 30-100 cm long and up to 60 cm wide. Pinnules triangular to lanceolate, lobed. Sori forming a continuous band along the margins of the segments on the lower parts of the lobes and protected by the reflexed margin of the pinna. (**38.** x0.5)

Common in drier forest, particularly coastal, in the North Island and in the north and west of the South Island as far south as Greymouth.

39. Pteris tremula. Rhizome short, erect. Fronds tufted, erect, 30-150 cm long. Lamina bright green, soft, membranous. Sori forming a band or broken line along the margins of the pinnules. (**39.** x0.3)

Quite common in the forests of the North Island particularly in drier places. It occurs in Nelson, Marlborough and Banks Peninsula in the South Island.

40. Athyrium japonicum. Rhizome creeping, covered with brown scales. Fronds lanceolate, 15-50 cm long, light to dark green, soft. Sori oblique to midrib, oblong, covered by a delicate indusium (40a). (**40.** x0.4; **40a.** x1.3)

Restricted to the North Island and extending from Kaitaia to Taumarunui but rare south of Hamilton. It is quite common in Northland under open forest and along tracks; appears to be actively spreading.

41. Hypolepis rufobarbata. Rhizome long creeping, covered with reddish-brown hairs. Fronds lanceolate usually 20-80 cm long, 10-25 cm wide. Stipe clothed in reddish-brown glandular and non-glandular hairs of varying lengths. Lamina including the rachis clothed in soft reddish-brown hairs. Many of the hairs, particularly on the underside, glandular (41b). Sori (41a) 1-6 per pinnule, orange-brown, often covering the pinnule when old. (**41.** x0.2; **41a.** x20; **41b.** x3)

Abundant throughout the country in open forests, forest clearings, forest margins and on sheltered banks.

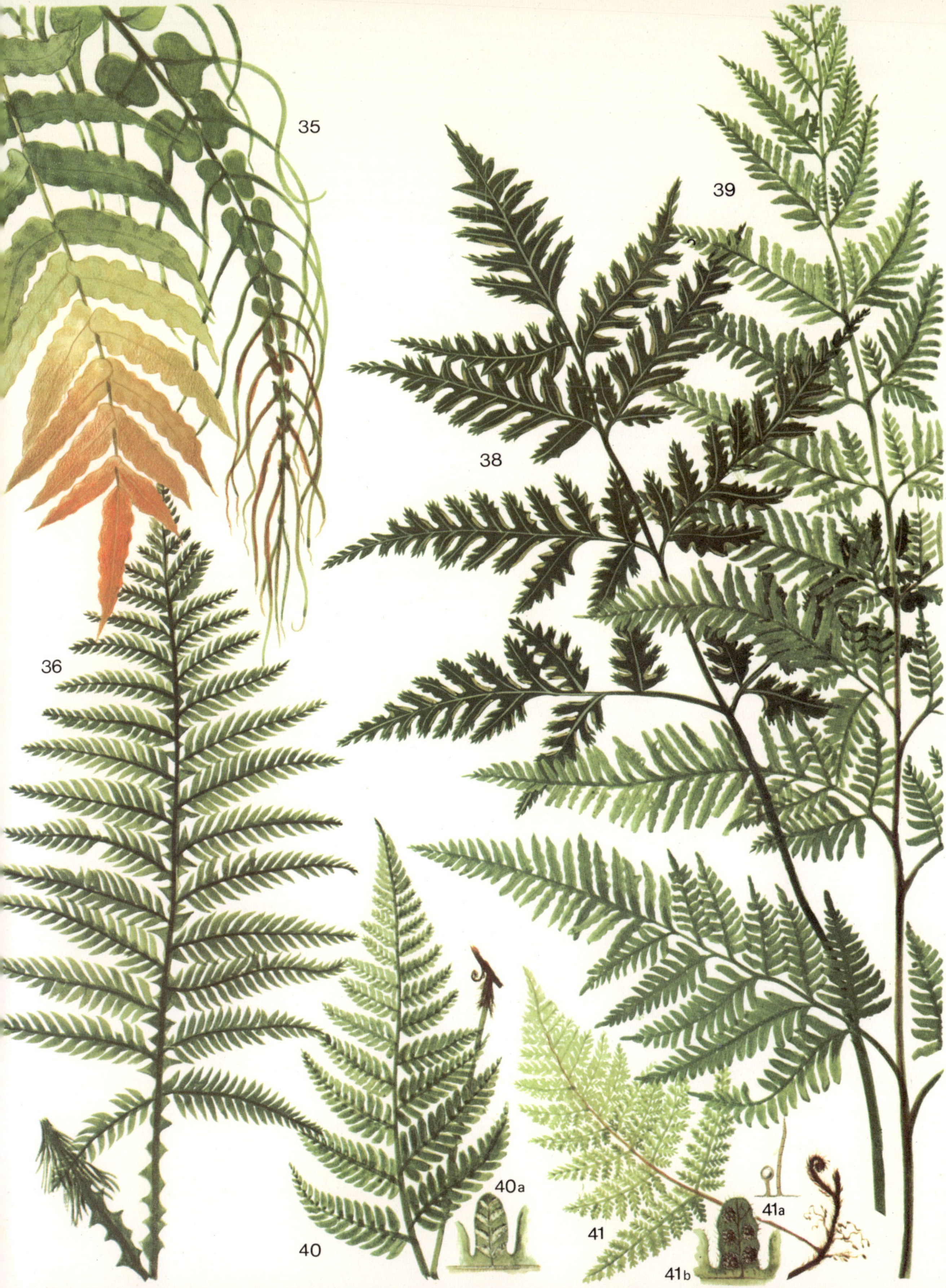
35
39
38
36
40a
41a
40
41
41b

BRACKEN and BRACKEN-LIKE FERNS

42. Histiopteris incisa. "Histiopteris". Rhizome thick, scaly, long creeping. Fronds glabrous, broad lanceolate or lanceolate, 20-100 cm or more long, green to bluish-green. Young uncoiling fronds a distinctive milky bluish-green. Pinnae opposite or sub-opposite, pinnules or fertile fronds narrower. Sori forming a continuous band along the margins of the pinnules but not reaching the tip. Indusium thin, extending from the margin (42a). (**42.** x0.7; **42a.** x1)

An abundant fern throughout the country. It thrives in open situations along the margins of forest, in forest clearings, along streams, thermal areas and in the mountains where it often forms thickets along roadsides. Old fronds die off during autumn and new fronds start developing from September onwards.

43. Pteridium esculentum. "Bracken". Rhizome much branched, subterranean, black, fleshy. Fronds normally 1-2 m high but in very shaded places may be up to 4 m high. During uncoiling the rachis rotates and the lower pinnae pairs project at different angles. Pinnae very firm, harsh, curved, green to brown. Sori forming a continuous band along the margins of the narrow segments and protected by the reflexed margin of the segment. Very old fronds are completely covered on their undersides by old sporangia. (**43.** x1 upper portion of frond)

Abundant to over-abundant throughout the country. Very common along the margins of forest, forest clearings, along the edges of roads, scrublands and mountain regions. A pioneer fern that often invades land after forest or scrub is cleared.

44. Paesia scaberula. "Ring Fern" or "Hard Fern". Rhizome creeping, wiry, covered with brown scales. Fronds 25-60 cm long, often drooping, at first sticky due to short glandular hairs. Rachis distinctly zig-zag, rough. Lamina green to yellowish-green, tripinnate, covered with short glandular hairs. Sori forming a band along the edges of the pinnules (44a) but not reaching the base or tip. Sorus protected by the thin pinnule margin and a delicate indusium. (**44.** x0.7; **44a.** x1)

Abundant throughout the country, particularly on open sunny hillsides where it often covers extensive areas. Fronds turn reddish-brown and die off in late autumn or earlier if frosts are prevalent.

45. Hypolepis tenuifolia. Rhizome long creeping, much branched, covered with dark brown hairs. Fronds very variable, lanceolate to broad lanceolate to deltoid 20-150 cm or more long. Stipe yellowish-brown to dark brown, hairy or glabrous, rough in lower part. Lamina light to dark green, glabrous or hairy. Hairs of rachis and midribs whitish to brown, septate, gradually tapering. Lowermost pinnae oblique to rachis, upper ones almost at right angles to it. Sori 2-8 per ultimate segment usually unprotected (45a) but occasionally with a weakly developed reflexed flap extending from the margin. (**45.** x0.7; **45a.** x10; **45b.** x20)

Abundant throughout the country except in Northland where it is rare. Usually found on the margins of forest, forest clearings, under open forest or on hillsides. Often found growing with other brackens. Very hairy forms of the plant are encountered in colder regions.

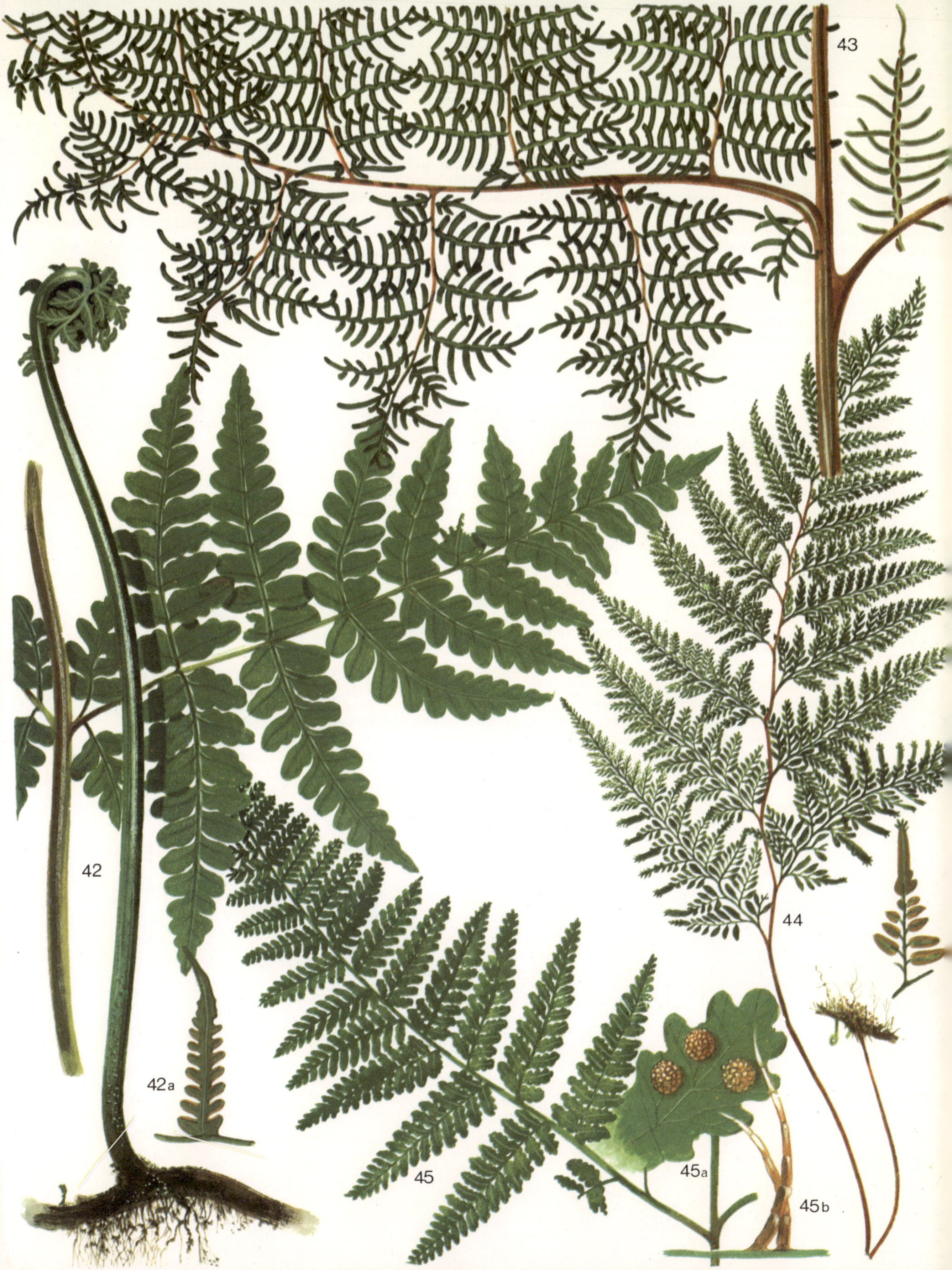
43
42
44
42a
45
45a
45b

TREE-FERNS

The tree-ferns form a conspicuous part of our flora and their slender trunks and graceful spreading crowns are known to everyone. Eight species occur on our main islands and of these Mamaku (*Cyathea medullaris*) and Ponga (*Cyathea dealbata*) are perhaps the best known.

There are two genera of tree-ferns in New Zealand, *Cyathea* and *Dicksonia*, and they are distinguished as follows: *Cyathea* has sori situated away from the margin and possesses both scales and hairs; *Dicksonia* has marginal sori and hairs only.

Often tree-ferns are too tall for us to examine the fronds and soral structures used in their identification, however, all the species can be readily identified in the field by their general habit and this is emphasised here in the descriptions. All habit illustrations are approximately one fiftieth natural size (0.02).

46. Cyathea medullaris. "Mamaku", "Black tree-fern". Trunks to 15 m high, black, with a distinct hexagonal leaf-scar pattern, slender, but expanded near the base by a mat of aerial rootlets. Fronds large, spreading, pinnae drooping. Stipe black, rough, scales blackish-brown, cells orientated in same direction, margin toothed (46b). Lamina dark green, tripinnate, thick. Sori large, dark brown, covered with an indusium (46a). (**46a.** ×10; **46b.** ×1)

Abundant throughout the country except in some drier east coast districts. Mamaku is particularly common along water courses in lowland forest. It has also established well in exotic forests near Rotorua.

47. Cyathea smithii. Trunks 3-8 m high, the upper part covered by old frond stalks. Fronds long and narrow, bright green, thin and membranous. Stipe scales (47a) brown, cells near margin oriented in different directions. Sori partially covered by a scale-like indusium (47b). (**47a.** ×1.2; **47b.** ×5)

Found from Kaitaia southwards particularly in hilly forest valleys. More abundant in mountain regions of the North Island. A small tree-fern which is easily identified by its skirt of old frond stalks.

48. Cyathea dealbata. "Ponga", "Silver Fern". Trunks 2-10 m high, stout, leaf bases persistent. Fronds erect, spreading. Stipe 2-4 cm thick near the base, rough, covered with shining chestnut-brown scales. Cells near the margin of the scale oriented in different directions. Lamina bright green above, white to light grey below (except in very young plants). Sori large, brown, covered with an indusium.

Illustration (on back cover) shows pinna at left (×0.5), scale, above centre (×5) small portion much enlarged, uncoiling frond (×0.7), secondary pinna bottom centre (×4), and the tree-fern habit (×0.02).

Abundant throughout the country. In forests it is particularly common on drier slopes and ridges. It also thrives in open scrub particularly after forest has been cleared. Ponga is easily identified by the white to grey underside of the lamina.

49. Dicksonia fibrosa. "Wheki-ponga". Trunks up to 7 m high, reddish-brown, very soft due to a complete outer covering of aerial rootlets, stout, up to 50 cm diameter. Fronds dark-green, firm, numerous. Stipe and rachis smooth, covered at first with soft brown hairs. Sori numerous, marginal, protected by the modified segment margin and a delicate semi-transparent indusium (49a). (**49a.** ×2.5)

Found from Auckland southwards in forest but often rare or absent for large areas. Easily recognised by its stout soft trunk and skirt of old fronds that persist for years.

50. Dicksonia squarrosa. "Wheki". Trunks slender, blackish-brown, very rough due to persistent stipe bases. Main trunk producing buds and stolons giving rise to new trunks. Fronds small, harsh, stipe and rachis very rough. Sori similar to the above.

Abundant throughout the country in forests and open places on poorly-drained soils and in swamps. Wheki is easily recognised by its numerous trunks and very rough stipe, rachis and pinnae midribs.

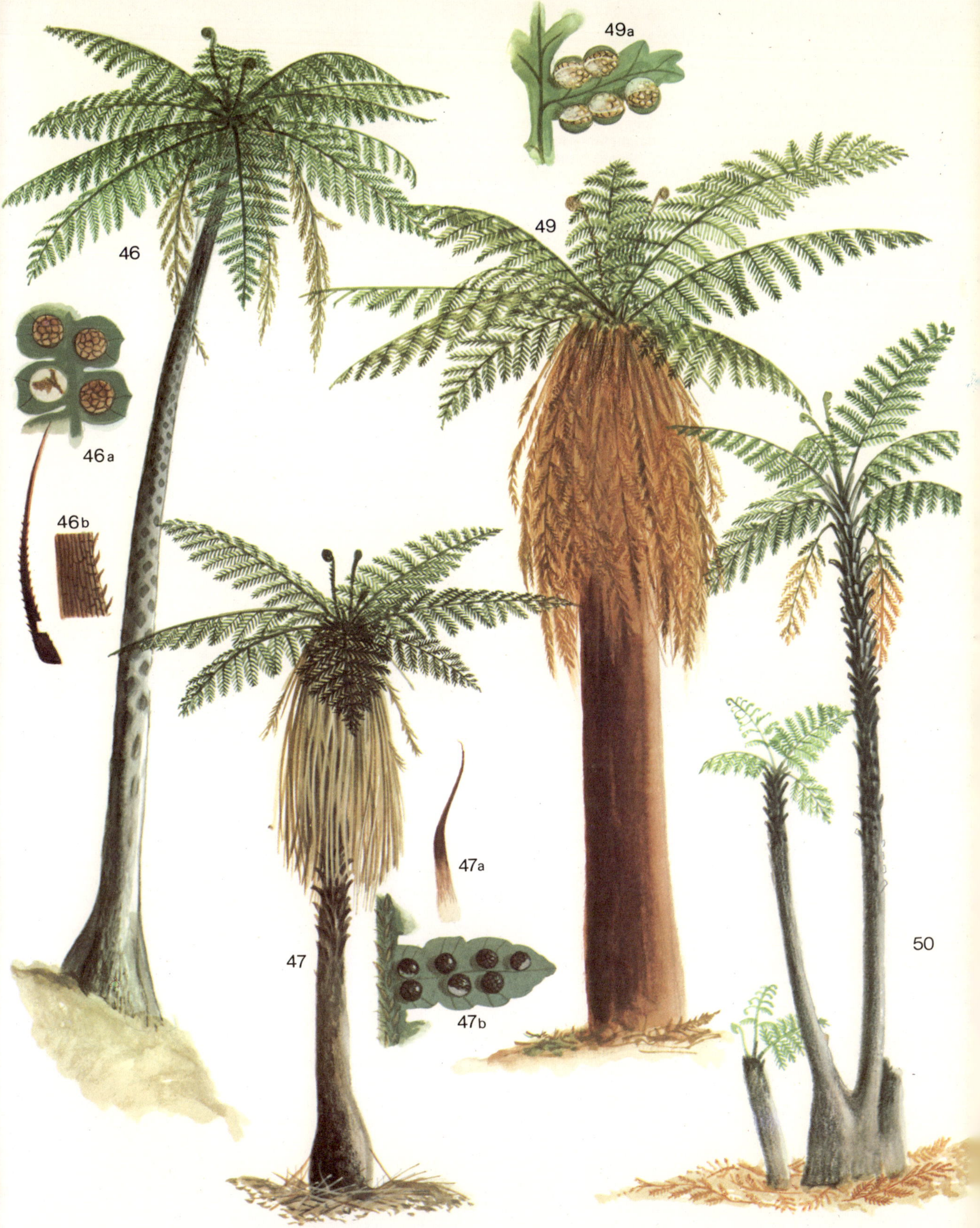
49a
46
49
46a
46b
47a
47
47b
50

FILMY-FERNS

The Filmy-ferns comprise almost one fifth of our fern flora. Their fronds are very thin and delicate and most of the species thrive only in the moist forest environment. At first glance many filmy-ferns look alike but with the aid of a hand lens their differences are soon made clear. It is not uncommon to find 10 or more species growing close together. Some classifications break the filmy-ferns up into a large number of genera but *Hymenophyllum* and *Trichomanes* are used here although the alternative generic name for each species is given in brackets.

51. Hymenophyllum *(Mecodium)* **flexuosum.** Fronds erect or drooping, dark green. Stipe and rachis with a flexuous green wing. Pinnae segments often flexuose. Sori at the ends of segments, valves broad, rounded. (**51.** x0.6)

Commonly encountered along stream courses on banks or rocks.

52. Hymenophyllum *(Mecodium)* **dilatatum.** Fronds 10-50 cm long, lamina bright green. Segments of pinnae very broad, 2-3 mm. Stipe smooth. Sori (52a) very large, terminating segments. (**52.** x0.8; **52a.** x3)

Abundant throughout the country in forest as a epiphyte or growing on logs or banks. Easily identified by the broad pinnae segments.

53. Hymenophyllum *(Mecodium)* **sanguinolentum.** Fronds olive-green, glabrous, up to 20 cm long, often very slender in the upper part. Stipe hairy near base. Rachis, midribs of pinnae and smaller veins black. Sori at the ends of segments, valves with claws on their backs. (**53.** x0.5)

Abundant throughout the country in lowland forest as an epiphyte or on the forest floor. When dried this species emits a peculiar odour similar to dried blood.

54. Hymenophyllum *(Mecodium)* **demissum.** Fronds erect or drooping, bright green. Stipe smooth, rachis winged, the wing occasionally wavy. Sori at the ends of segments, usually paired, and often partially fused together (54a). (**54.** x0.8; **54a.** x5)

Our most common filmy-fern, abundant everywhere. A ground-growing species that often covers extensive areas of the forest floor. Occasionally epiphytic.

55. Hymenophyllum *(Mecodium)* **scabrum.** Fronds up to 60 cm long, deep green to brownish-green. Stipe covered with stiff brown hairs, often glabrous on old fronds. Rachis and midribs of pinnae hairy. (**55.** x0.5)

An epiphytic species found from Kaitaia southwards but common only south of Hamilton.

56. Hymenophyllum *(Meringium)* **multifidum.** Fronds bright to dark green, 2-30 cm long, pendulous or erect, frond often curved. Pinnae segments finely serrate. Sori large, projecting above the frond, valves to 3 mm long. (**56.** x0.3)

The most tolerant of our filmy-ferns common in the forest as an epiphyte or ground fern but also at home under sub-alpine scrub and in rock crevices where it is often dwarfed.

57. Hymenophyllum *(Mecodium)* **flabellatum.** Rhizome and stipe at first covered with brown woolly hairs. Fronds 4-25 cm long, light green, pinnae fan-shaped. Sori small, at the ends of segments (57a). (**57.** x0.5; **57a.** x4)

An attractive species with distinct fan-shaped pinnae. Widespread in forest throughout the country.

58. Hymenophyllum cupressiforme. Fronds 2-5 cm long. Rachis winged throughout, segments toothed. Sori large, projecting above the frond, valves entire or with slightly wavy margins. (**58.** x1; **58a.** x4)

A ground- or rock-growing species found in scattered localities in Northland, Auckland, Wairarapa and Wellington. A common Australian species that has only recently been recognised in this country.

59. Hymenophyllum revolutum. Similar to the above species but the rachis is not winged in the lower part and the valves of the sorus are deeply toothed. (**59.** x1; **59a.** x4)

An abundant little forest fern found growing on trees and fallen logs.

60. Hymenophyllum peltatum. Fronds 2-10 cm long, pinnae divided on the upper side only. Sori usually 2-3 together close to the rachis. (**60.** x1)

A small epiphytic mountain fern found from the Kaimai Range southwards.

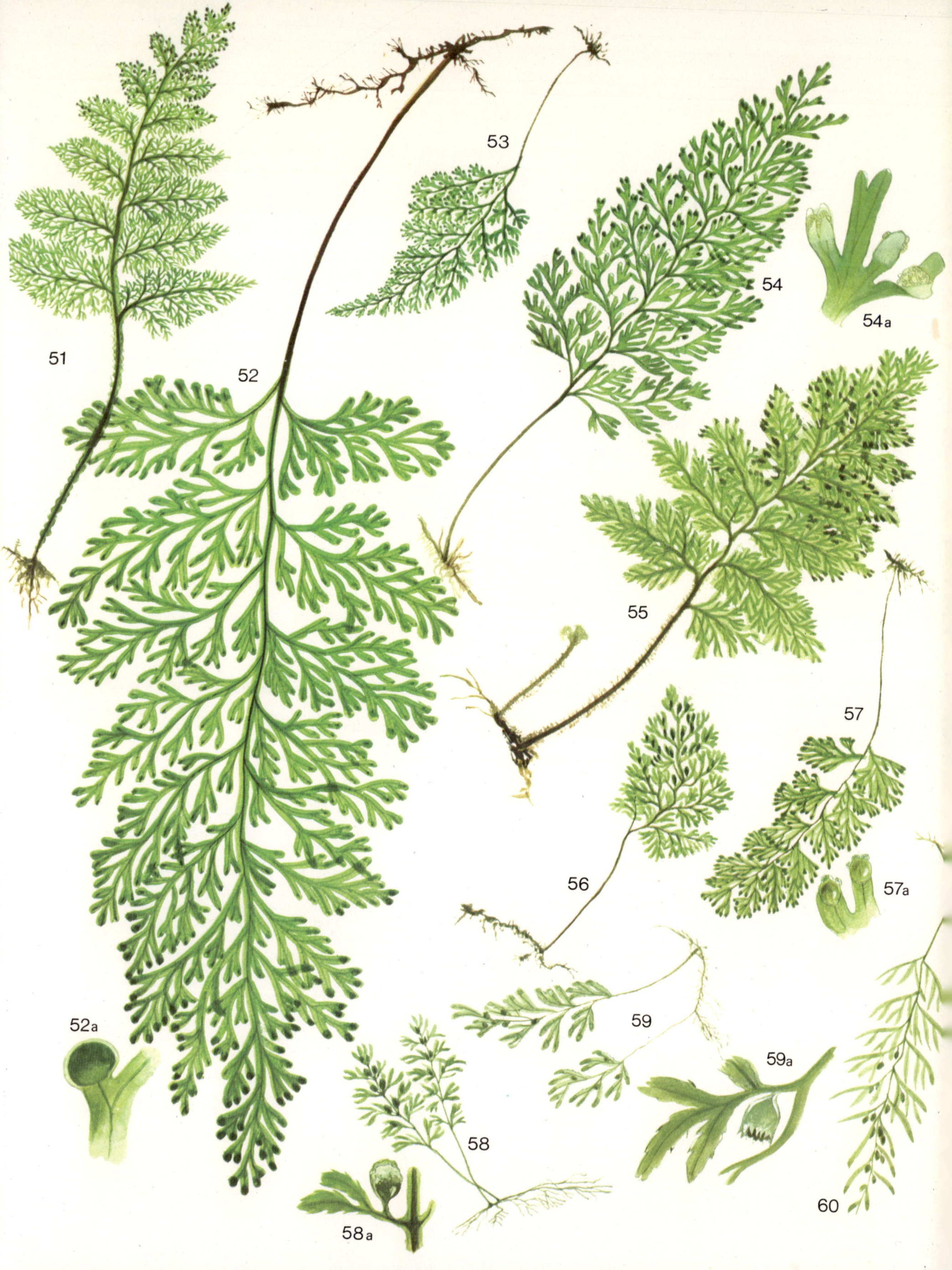

51
52
52a
53
54
54a
55
56
57
57a
58
58a
59
59a
60

FILMY-FERNS

61. Hymenophyllum *(Sphaerocionium)* **ferrugineum.** Fronds 5-20 cm long, pendulous, lightly to densely clad in reddish-brown stellate hairs (61a). Sori small immersed in the tips of segments. (**61.** ×0.7; **61a.** ×20)

Found from Kaitaia southwards. Usually growing on tree-ferns in wet valleys along stream courses. Easily identified by its reddish-brown fronds.

62. Hymenophyllum *(Apteropteris)* **malingii.** Fronds 4-16 cm long, very narrow. Stipe filiform, lamina reddish-brown to grey due to a dense covering of stellate hairs. (**62.** ×0.6)

A very distinct mountain species almost entirely found on mountain cedar (*Libocedrus bidwillii*) and occurring from Mt Te Aroha southwards.

63. Trichomanes *(Cardiomanes)* **reniforme.** "Kidney-fern". Easily identified by its characteristic kidney shaped fronds. (**63.** ×0.7)

Common in most parts of the country except in the east coast of the South Island. Kidney-fern often forms extensive mats on the forest floor on banks and on rocks. Occasionally epiphytic.

64. Trichomanes *(Polyphlebium)* **venosum.** Fronds 2-8 cm long, pendulous, light green, translucent. Sori immersed in lobes of segments near the rachis, bristle up to 10 mm long (64a). (**64.** ×1; **64a.** ×5)

Abundant in forests throughout the country, particularly on the trunks of tree-ferns.

65. Trichomanes *(Crepidopteris)* **endlicherianum.** Fronds 4-9 cm long, very dark green. Sori immersed in the tips of lateral segments, bristle to 10 mm long. (**65.** ×1.5)

Common in the North Island and in the north and west of the South Island to latitude 44°. A small filmy fern usually found on rocks and banks near streams. Often overlooked.

66. Trichomanes *(Selenodesmium)* **elongatum.** Fronds tufted, broadly triangular, very deep green. Sori in sinuses of pinnule lobes, bristle projecting from the underside and up to 15 mm long. (**66.** ×0.8)

Very common on wet banks near streams from Northland to East Cape. Further south it is rare but extends into Nelson, Marlborough and Canterbury. Easily identified by its deep green triangular fronds.

67. Trichomanes *(Vandeboschia)* **colensoi.** Fronds 4-12 cm long, dark green, finely cut. Sorus tubular, near rachis at base of pinna (67a). (**67.** ×1; **67a.** ×5)

Found from Rotorua southwards but more common in the South Island. This species is usually found on rocks and banks in very dark places and is often associated with *T. endlicherianum.*

68. Hymenophyllum *(Mecodium)* **rarum.** Fronds usually 3-15 cm long, pale greyish to milky green, narrow oblong. Sori immersed in the tips of segments, valves rounded. (**68.** ×0.8)

Particularly common in mountain areas of the North Island but more widespread in the South Island. Common as an epiphyte or on rocks.

69. Hymenophyllum *(Sphaerocionium)* **lyallii.** Fronds small, fan-like, pale green. Lamina segments minutely toothed, sori immersed in tips of segments (69a). (**69.** ×1; **69a.** ×4)

This species extends from Northland southwards but it is often rare and local. Usually found as an epiphyte on tree-ferns but occasionally on banks and on rocks.

70. Hymenophyllum *(Craspedophyllum)* **armstrongii.** Fronds 1-2 cm long, simple or forked. Lamina toothed. Sori terminal, large, black. (**70.** at left, ×0.8, enlargement ×1.5)

A small filmy-fern epiphytic on trees or forming mats on rocks in mountain areas. Often overlooked because of its small size.

71. Hymenophyllum *(Meringium)* **minimum.** Fronds 2-8 mm long with 1-6 pairs of toothed segments. Sori terminal, large, valves with spiny margin and back. (**71.** at left, ×1; enlargement ×3)

Found throughout the South Island and Stewart Island in open rocky places. In the North Island it is known only from the Wellington coast and in the Wairarapa. Often forms carpets on rocks.

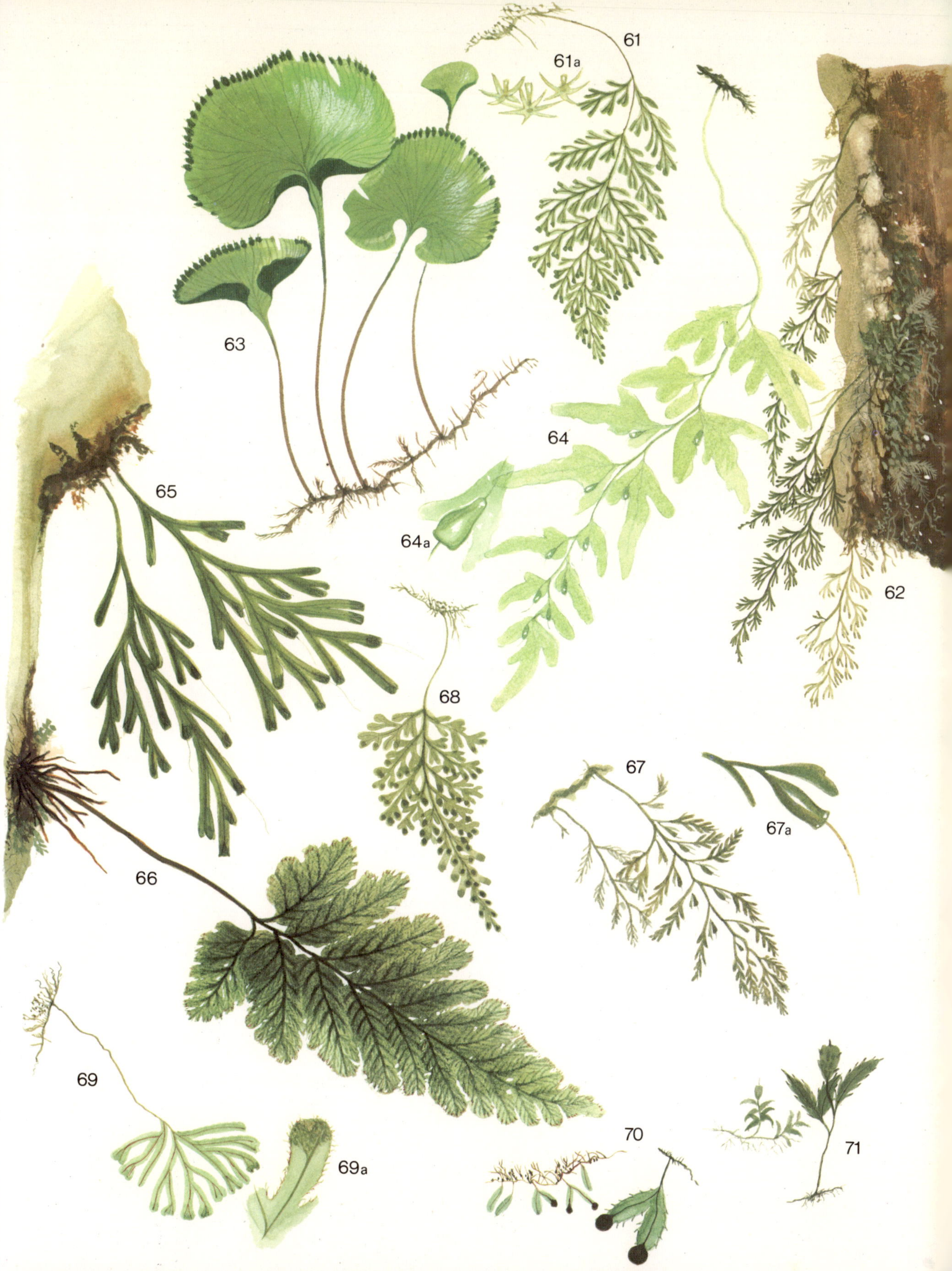
61
61a
62
63
64
64a
65
66
67
67a
68
69
69a
70
71

CRAPE FERN
and others

72. Cyclosorus penniger. Rhizome short erect, or forming a small trunk. Fronds 30-150 cm long, lanceolate. Stipe dark brown, grooved, scaly near the base. Rachis brown, often pubescent at first, grooved above. Pinnae bright green, membranous, deeply lobed. Sori in two rows near the midrib, naked. Mature sporangia black. (**72.** ×0.5; **72a.** ×1)

An abundant forest fern found throughout the North and South Islands. This species thrives on moist alluvial soils and is particularly common in forest valleys and along water courses.

73. Lindsaea trichomanoides. Rhizome slow creeping. Fronds 8-30 cm long. Stipe slender, wiry, reddish-brown, glabrous. Lamina pinnate, bipinnate, or sub-tripinnate, lanceolate, thick. Pinnae variously dissected. Sori forming a band along the apex of the lobe. Indusium broad, delicate (73a). (**73.** ×0.5; enlargement ×3)

Common throughout the North Island but rare and local in the South Island. A particularly common species in Beech (*Nothofagus spp.*) and Kauri *(Agathis australis)* forests on drier banks or at the base of trees. From Rotorua northwards the most commonly encountered form has less divided fronds with larger final segments and it is distinguished as a distinct variety (*Lindsaea trichomanoides* var. *lessonii).*

74. Lindsaea linearis. Rhizome creeping, scaly. Fronds 10-25 cm long, pinnate. Sterile fronds deep green, spreading, pinnae fan-shaped, 4-6 mm wide. Fertile fronds erect, pinnae narrow, often brown. Sori extending along the upper margin of the pinna. Indusium delicate, with an irregular margin. (**74.** ×0.5)

This species is usually found on clay and other poorly-drained soils under scrub. Very common as far south as East Cape. Further south it is uncommon but extends to Stewart Island.

75. Leptopteris superba. "Heruheru", "Crape Fern", "Prince of Wales Feather". Rhizome erect, often forming a trunk up to 1 m high. Fronds 20-100 cm or more long, rich deep green, lanceolate, clustered together and forming a crown, lowermost pinnae gradually tapering to the stipe. Pinnae feathery, finely dissected, the pinnule segments overlapping. Sporangia large, light brown, scattered along the midribs of the pinnule segments in the lower half of the pinna. (**75.** ×0.5)

This species thrives only in cool very moist forests and attains its greatest size in the western and southern parts of the South Island. In the North Island it is predominantly a mountain forest fern. It is rare north of Rotorua but has recently been re-discovered in the Waipoua Kauri Forest, Northland.

76. Leptopteris hymenophylloides. "Heruheru". Rhizome short, erect often forming a short trunk. Fronds crowded, up to 90 cm long, deep green. Stipe green to brown, winged at base. Lamina finely dissected, pinnae spreading, segments flat, semi-transparent. Sporangia large (76a), light brown, scattered along the midribs of the pinnule segments towards the lower part of the pinna. (**76.** ×0.5; **76a.** ×20; **76b.** ×5)

Abundant throughout the country in lowland to montane forests particularly along stream courses and damp banks. Young plants are easily mistaken for a filmy-fern.

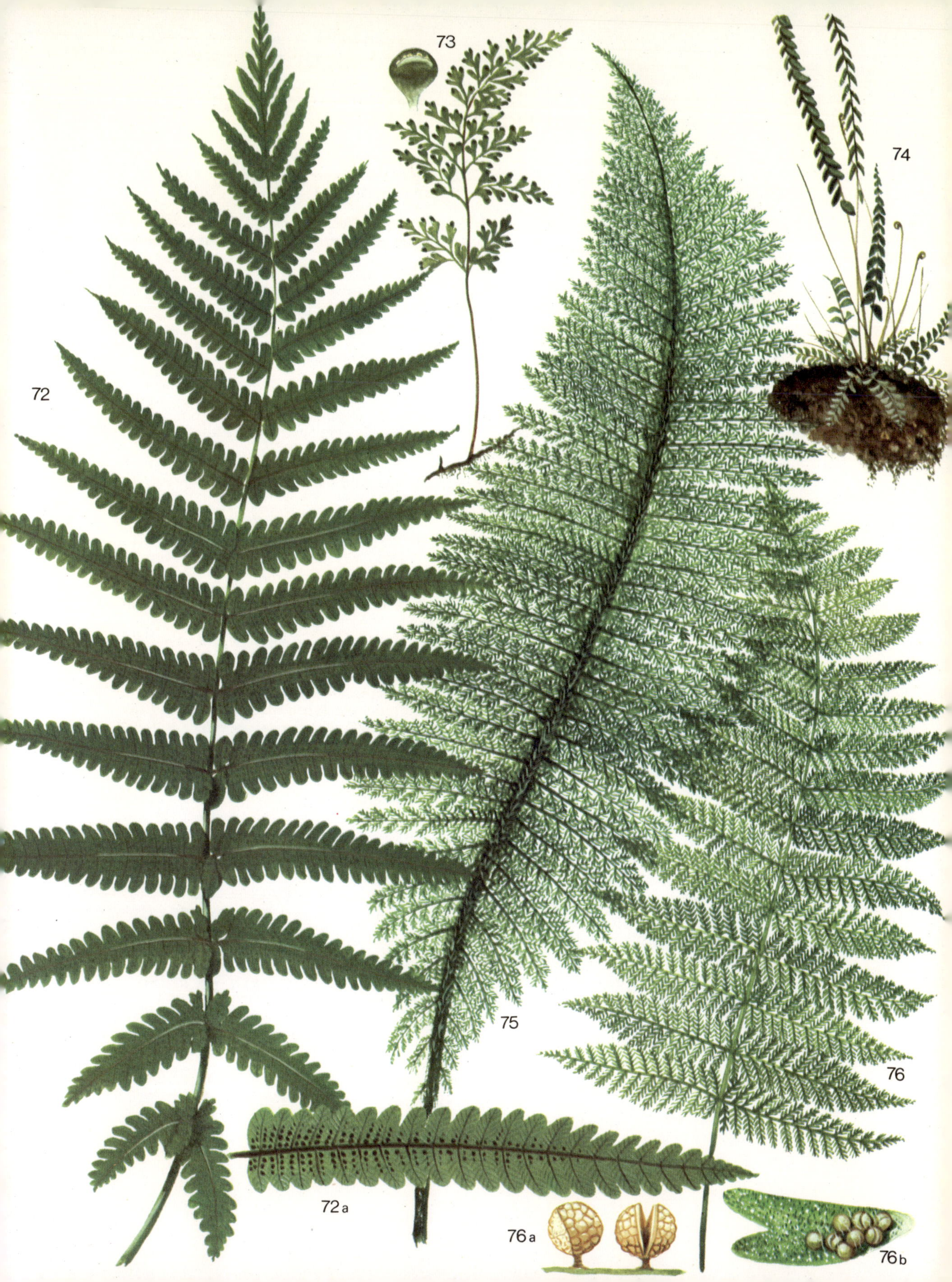
73
74
72
75
76
72a
76a
76b

UMBRELLA FERNS and LADDER FERN

77. Gleichenia dicarpa. *(Gliechenia circinata).* "Swamp Umbrella Fern". Rhizome slender, wiry, clad with scales. Fronds variable, 20-100 cm long and consisting of 1-3 or more tiers of pinnae. Pinnae forked 2-3 times, final segments 2-3 mm long, rounded and pouched on the underside, with 1-2 brown sporangia. (**77.** x0.3)

Common throughout the country in open scrublands where it often forms dense thickets. Particularly common Northland. Another very closely related species which often grows in similar situations is *G. microphylla*. This species is best distinguished by its triangular flat final segments and its 2-4 bright yellow sporangia per segment.

78. Gleichenia dicarpa var. **alpina.** Fronds small, rarely more than 30 cm high, greyish-green, crowded together. Rachis and undersides of pinnae covered with reddish brown scales and hairs. (**78.** x0.8)

A common little mountain fern extending from Mt Moehau, Coromandel Peninsula, southwards. It thrives in mountain swamps and usually forms dense cushions. In the southern South Island it descends to sea level. It is possible that this variety is merely a dwarfed form of the species.

79. Gleichenia cunninghamii. "Umbrella Fern". Rhizome creeping, covered with dark brown scales. Stipe 10-20 cm long, covered at first with brown scales and hairs. Lamina bright green above, whitish below, divided into two primary pinnae (as illustrated), with normally a dormant bud in the fork between the pinnae. This bud may develop and produce 1 or 2 accessory tiers of pinnae. Sori copious, in two rows on either side of the midrib of the final segments. (**79.** x0.5; enlargement x5)

Common throughout the North Island, particularly at higher altitudes, less common and local in the South Island. Usually encountered on banks or in more open forest. Ground forms usually develop a number of pinnae pairs.

80. Gleichenia flabellata. Rhizome stout, covered with reddish-brown scales, usually subterranean. Fronds 30-80 cm long, deep green. Stipe stiff, smooth, dark brown. Lamina divided into two primary pinnae (illustrated) or with 2-3 pairs, segments oblique to midrib. Sori numerous, in two rows, consisting of 1-5 yellow sporangia. (**80.** x0.4; **80a.** x3)

Restricted to the Northern North Island from North Cape to Auckland and the Coromandel Peninsula. It is known from near Takaka, NW Nelson in the South Island. It is quite a common fern in Northland and thrives in open scrubland areas, along streams and on banks.

81. Nephrolepis cordifolia. "Ladder Fern" or "Sword Fern". Rhizome short, semi-erect, producing numerous scaly wiry stolons which eventually root and produce new plants. Fronds tufted, 30-70 cm long pinnate. Rachis covered with scales. Pinnae close together, lowermost small and sterile. Sori in two rows along the pinna, usually closer to the margin. Indusium kidney-shaped. (**81.** x0.5; **81a.** x3)

This species is restricted to the thermal areas between Rotorua and Taupo where it thrives on banks and cliffs near thermal pools. Another adventive species of Ladder Fern is particularly common around Auckland and other centres but it can be easily distinguished by its larger more erect fronds and the presence of small potato-like tubers, 1-2 cm diameter, on its roots.

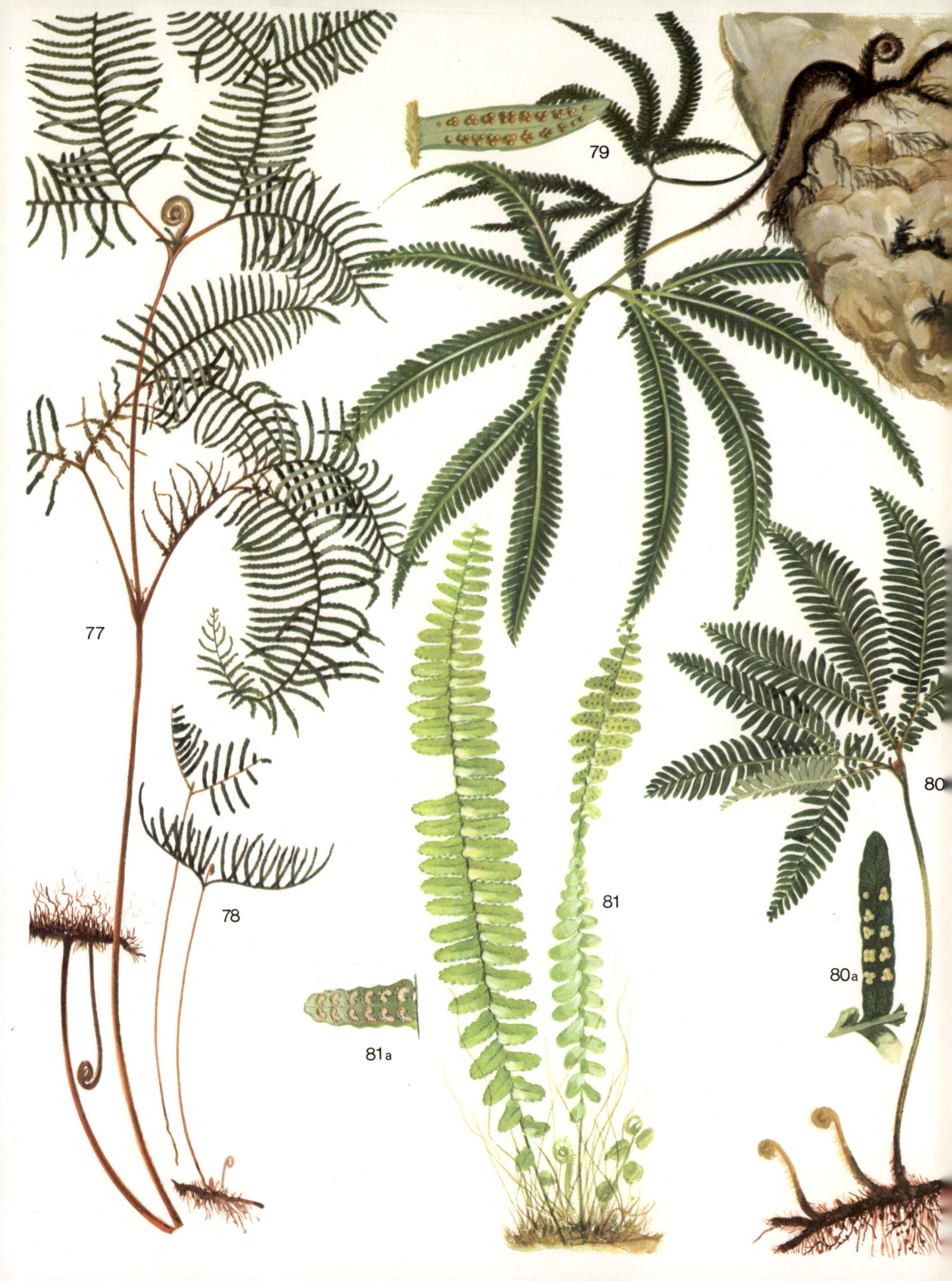
77
78
79
80
80a
81
81a

PARSLEY FERN, COMB FERNS
and others

82. Botrychium australe. "Parsley Fern". Rhizome short, erect, roots thick, fleshy. Sterile frond fleshy, green or bronze, triangular. Ultimate divisions 3-5 mm wide. Fertile spike erect, projecting above sterile frond, pinnae narrow, folded. Sporangia large, brown when mature, partially protected by folded pinnae (82a). (**82.** x1; **82a.** x6)

Found in many scattered localities throughout the North and South Island. Usually found on the margins of forest or along tracks and in clearings. Often overlooked.

83. Ophioglossum coriaceum. "Adder's Tongue Fern". Rhizome erect, tuberous. Sterile fronds bright green, variable in shape but usually lanceolate, 1-8 cm long, fleshy. Fertile spike projecting above sterile frond or almost sessile. Stalk fleshy, white. Sporangia in two rows, embedded in the spike (83a). (**83.** x1.5; **83a.** x10)

Widespread throughout the country in open grasslands and along river flats. A small species which is easily overlooked.

84. Schizaea fistulosa. "Comb Fern". Rhizome slow creeping, covered with glossy dark brown hairs. Fronds clustered, 15-40 cm long. Stipe green, smooth, grass-like. Fertile lamina 2-3 cm long, pinnae folded, brown at maturity. (**84.** x1)

Very common from North Cape to East Cape on clay and poorly-drained soils, usually under scrub. It extends southwards from East Cape to Stewart Island but it is rare and local.

85. Schizaea bifida. Rhizome short, stout. Fronds clustered together, 6-30 cm long, simple or forked a few times. Stipe green, rough. Fertile lamina 1-2 cm long, 6-8 mm broad, light brown when mature. (**85.** x0.5)

Common from North Cape to East Cape on poor soils, usually under scrub and often associated with *S. fistulosa*. South of East Cape it is rare in the North Island. In the South Island it is confined to NW Nelson and is quite common on the pakahi soils between Takaka and Collingwood.

86. Schizaea fistulosa var. **australis.** Rhizome slow creeping, fronds crowded together, small, rarely more than 10 cm high. Fertile lamina 0.5-1.5 cm long. (**86.** x0.8)

Not uncommon in mountain swamps in the North Island but more widespread in the South Island and descending to sea level in the southern part and on Stewart Island. It is possible that this variety is merely a dwarfed form of the species.

87. Schizaea dichotoma. "Fan-fern". Rhizome short, erect. Fronds few, 10-35 cm long. Stipe brown near base, green above, stiff and wiry, divided dichotomously in its upper part into many broad spreading segments. Each segment terminating in a small stalked fertile lamina 3-5 mm long. (**87.** x0.5)

Fan-fern is restricted to the northern North Island from near North Cape to a little south of Rotorua. It was once very common in the Kauri forests but is now very rare and local. Usually found at the base of Kauri trees.

88. Lygodium articulatum. "Mangemange" (see page 6). Rhizome creeping, covered with brown hairs. Ground fronds (bottom right) divided into two pinnae, forked 2-4 times, pinnules up to 10 cm long. Climbing fronds of "unlimited growth", climbing over scrub and up trees 10-20 m or more. Sterile pinnae twice branched, pinnules green or yellowish-green. Fertile pinnae branched many times, pinnules small, deeply lobed 8-12 sporangia per lobe. (**88.** x0.7 appears on page 1).

Abundant in forests throughout the northern North Island to just south of Rotorua. Often forms thickets of numerous twining raches.

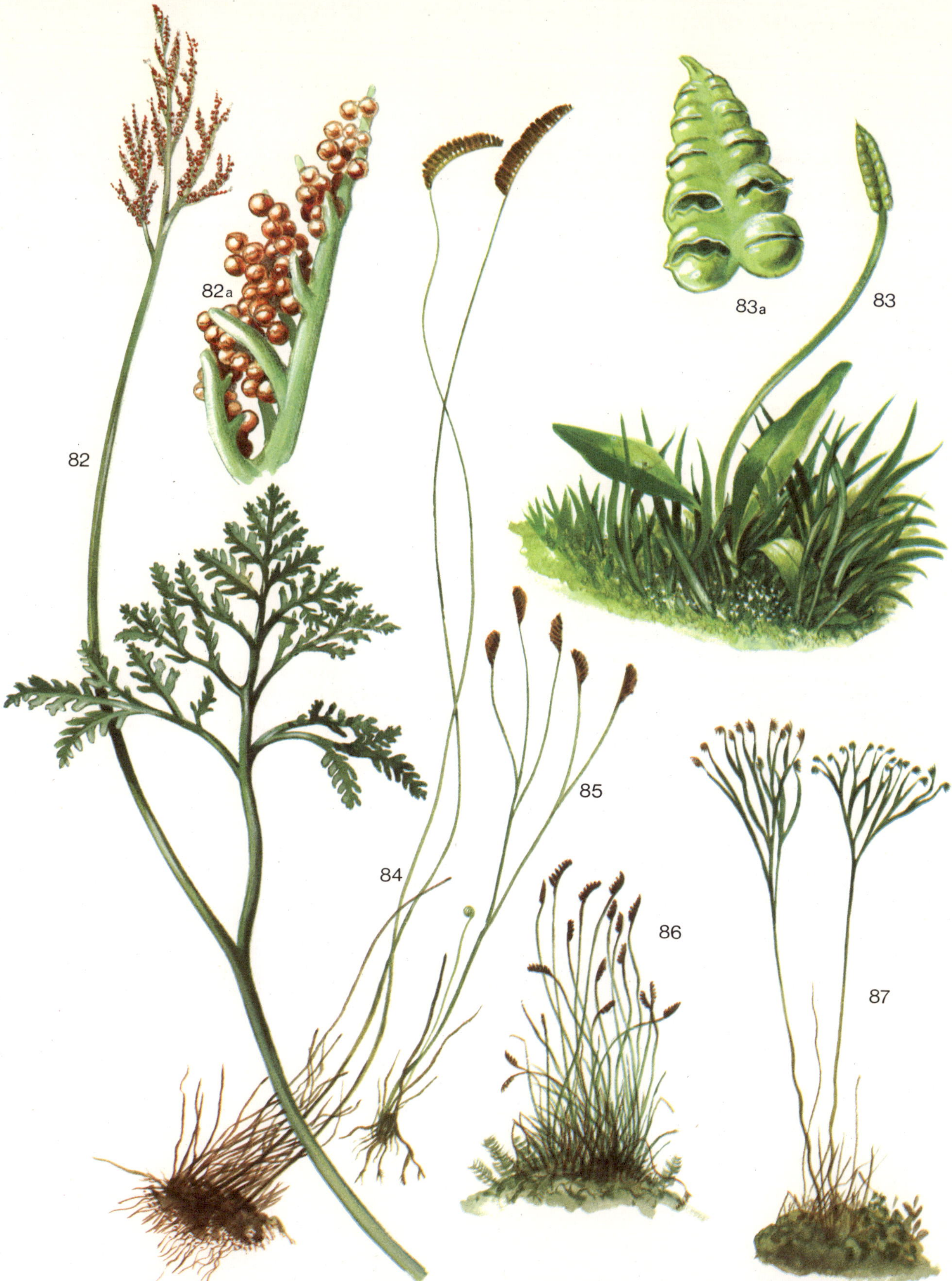
82
82a
83
83a
84
85
86
87

FERN ALLIES
LYCOPSIDA

The Lycopods are a much older group than the ferns and flourished during the Carboniferous era when they were the dominant vegetation. Many of the species were large woody trees and these are responsible for many of the coal measures that are used today.

Today's members are a far cry from those ancient giants and all the living species are small herbaceous plants rarely exceeding 50 cm high. *Lycopodium*, the largest genus, is widespread around the globe and the species are found in all situations from open swamps to forests. The stems have spirally arranged microphylls (sterile leaves) which are very small and often scale-like and sporophylls (fertile leaves) with sporangia at their bases, which may be similar to the microphylls or modified and aggregated together on a distinct structure called a strobilus.

Phylloglossum is a monotypic genus restricted to Australasia. The leaves and strobilus are produced annually and the plant persists from year to year by means of a small resting tuber.

Isoetes belongs to another ancient lycopod group that was also common in the Carboniferous era. The genus is widespread around the globe and the species are terrestrial, amphibious or aquatic. The leaves are long and grass-like and possess sporangia at their bases. There are two types of spores; microspores (male) which develop in the microsporangium, and megaspores, which are large and used in identifying species, develop in the megasporangium.

89. Lycopodium billardieri. Stems dichotomously branched many times and often forming dense clumps, pendulous. Microphylls spreading, 0.5-1 cm long, spirally arranged sporophylls 1-2 mm long, overlapping. The transition from microphylls to sporophylls is gradual. (**89.** x0.5)

A common epiphyte in forests throughout the country. Often associated with *Astelia* and *Collospermum* clumps on tree branches. Occasionally growing on the ground.

90. Lycopodium fastigiatum. Rhizome creeping and producing erect stems at intervals. Stems much branched, microphylls 3-5 mm long spirally arranged, green. In exposed mountain situations the whole plant is dwarfed and may be bright orange (90a). Strobili stalked, erect, solitary, paired or clustered together, yellow at maturity. (**90.** x0.7; **90a.** x1)

Common in montane to alpine grasslands and scrublands throughout the country. The species descends to lowland areas in the southern South Island.

91. Lycopodium australianum. Stems erect, branched, up to 30 cm high. Leaves spirally arranged, up to 9 mm long. Sporophylls similar to microphylls, developing in the upper parts, sporangia kidney-shaped, yellow (91a). Bulbils developed in clusters in the upper parts of the stem. (**91.** x0.5; enlargement x5)

Found from latitude 38° southwards in montane grasslands often under scrub. Exposed plants often turn orange.

92. Lycopodium laterale. Stems usually erect, simple or branched, up to 40 cm high. Sporophylls spirally arranged. Strobili on short lateral branches 0.5-2 cm long, brown when mature. (**92.** x0.5)

Found in swampy areas throughout the country but particularly common on the gumlands in Northland.

93. Lycopodium scariosum. Main stems creeping, freely branching. Aerial stems much branched, microphylls dimorphic, the upper ones flattened into one row, the smaller scale-like ones on the underside (93a). Strobili stalked, solitary or occasionally paired at the ends of branches. (**93.** x0.5; **93a.** x4)

Widespread throughout the country but very rare north of latitude 38°.

94. Lycopodium cernuum. Main stems creeping, branched, rooting at intervals. Aerial stems erect, up to 60 cm high, much branched. Microphylls small, narrow, incurved, crowded. Strobili numerous, at the tips of branches, drooping. (**94.** x1) Appears on title page.

Common on road cuttings, banks and thermal areas from North Cape to Lake Taupo. It occurs in the South Island from NW Nelson to just south of Westport.

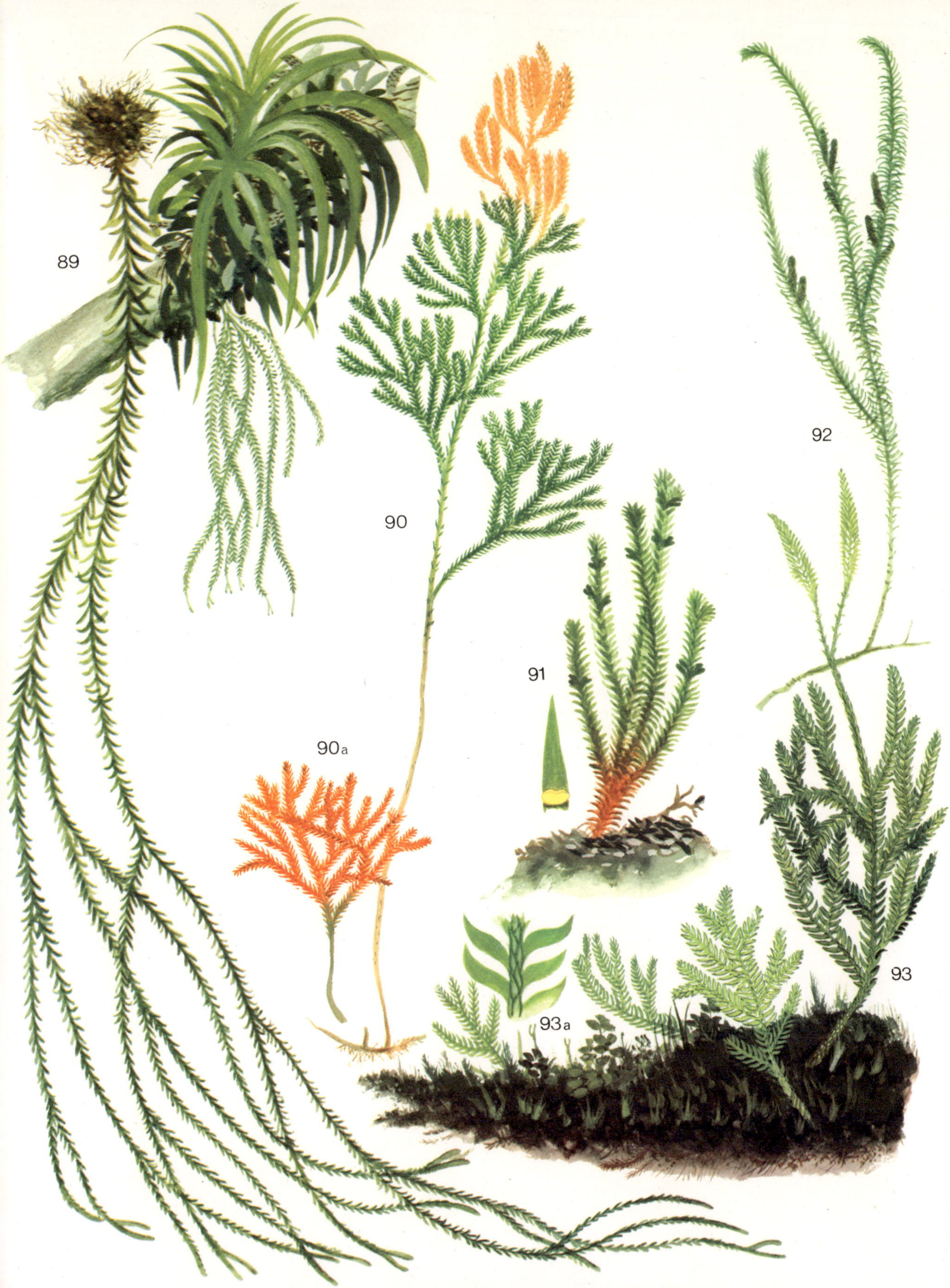
89
90
90a
91
92
93
93a

LYCOPSIDA

95. Lycopodium deuterodensum. Rhizome creeping, subterranean. Aerial stems robust, much branched, 30-100 cm long. Microphylls scale-like, appressed to stem or spreading. Strobili numerous, erect 1-3 cm long, yellow. (**95.** x0.5; enlargement x1)

A very robust species which is restricted to the North Island and extends from near North Cape to Lake Rotopounamu, just south of Lake Taupo. Usually found growing under scrub or on the margins of forest.

96. Lycopodium volubile. Main stems climbing, wiry, up to 5 m long. Microphylls dimorphic, the lateral ones flattened into one row, the upper ones small scale-like, appressed (96a). Strobili numerous, clustered together in large branched panicles. Sporophylls small, broad ovate, abruptly tapering to the tip (96b). (**96.** x0.8; **96a.** x5; **96b.** x10)

Common throughout the country in scrublands, on the margins of forest, and on banks and road-cuttings. In very exposed places the leaves may turn orange.

97. Lycopodium novae-zelandicum. Aerial stems erect or drooping, 5-20 cm long. Microphylls pale green, very soft and flexible, 8-12 mm long. Sporophylls similar to microphylls, towards the ends of branches, sporangium small, pale yellow (97a). (**97.** x0.6; enlargement x5)

A small forest species widespread throughout the country but never common. I have always found this species growing on tree-fern trunks. A series of intermediate forms link this species with *L. billardieri* (89) and it is possible that it is merely a habitat form of *L. billardieri.*

98. Lycopodium varium. Aerial stems erect, stiff, upper parts of branches drooping. Microphylls spreading, deep green, firm, 8-18 mm long. Strobili at the ends of branches, semi-erect or drooping. Sporophylls small, overlapping. (**98.** x0.5)

Found throughout the country but more common in the South Island and Stewart Island. In the North Island it is occasionally found in open exposed rocky mountain places. Always found growing amongst rocks under subalpine scrub or in more open forest situations.

99. Lycopodium serpentinum. Stems prostrate, creeping, rooting at intervals, usually tightly appressed to substrate. Microphylls linear, densely crowded. Strobili solitary, erect, on short lateral branches, up to 8 cm long. (**99.** x0.7)

Extremely rare in New Zealand and known only from Rukuhia near Hamilton and one locality on the Karikari Peninsula, Northland. This species grows in wet open swamps and soon vanishes when the swamp is drained. Very close to extinction in this country.

100. Lycopodium ramulosum. Aerial stems much branched, spreading, the tips of the branches ascending. Microphylls linear, spreading, green tinged reddish-brown in open situations. Strobili on short lateral branches, purplish-brown, 6-12 mm long. (**100.** x0.5; enlargement x2)

Restricted to poorly-drained soils in the western South Island and Stewart Island. Often forms extensive patches. In shaded places the plants grow more erect, often up to 20 cm tall.

101. Phylloglossum drummondii. Microphylls fleshy, 1-2 cm long, clustered together. Strobilus solitary, 4-8 mm long, stalk fleshy 1-3 cm long. New tuber (101a at right) developed each year. (**101, 101a.** x1)

Confined to the northern North Island and extending from North Cape to Auckland but rare in the southern part of the range. Quite common from Kaitaia northwards on open clay soils. Often forming colonies.

102. Isoetes kirkii. Plant aquatic, stem short, 3-4 lobed. Leaves 5-30 cm long, grass-like, usually all fertile. Sporangia at the base of the leaves, oblong 3-4 mm long. Megaspore (102a) white, when dry, 0.4-0.6 mm diameter, covered with tubercules. (**102.** x0.8; **102a.** x20)

Found from Northland (Wairua River) southwards to Lake Tekapo and surrounding tarns. Uncommon in the North Island. Our other species of *Isoetes, I. alpinus,* is similar in form to *I. kirkii*, but differs in having megaspores which lack tubercules.

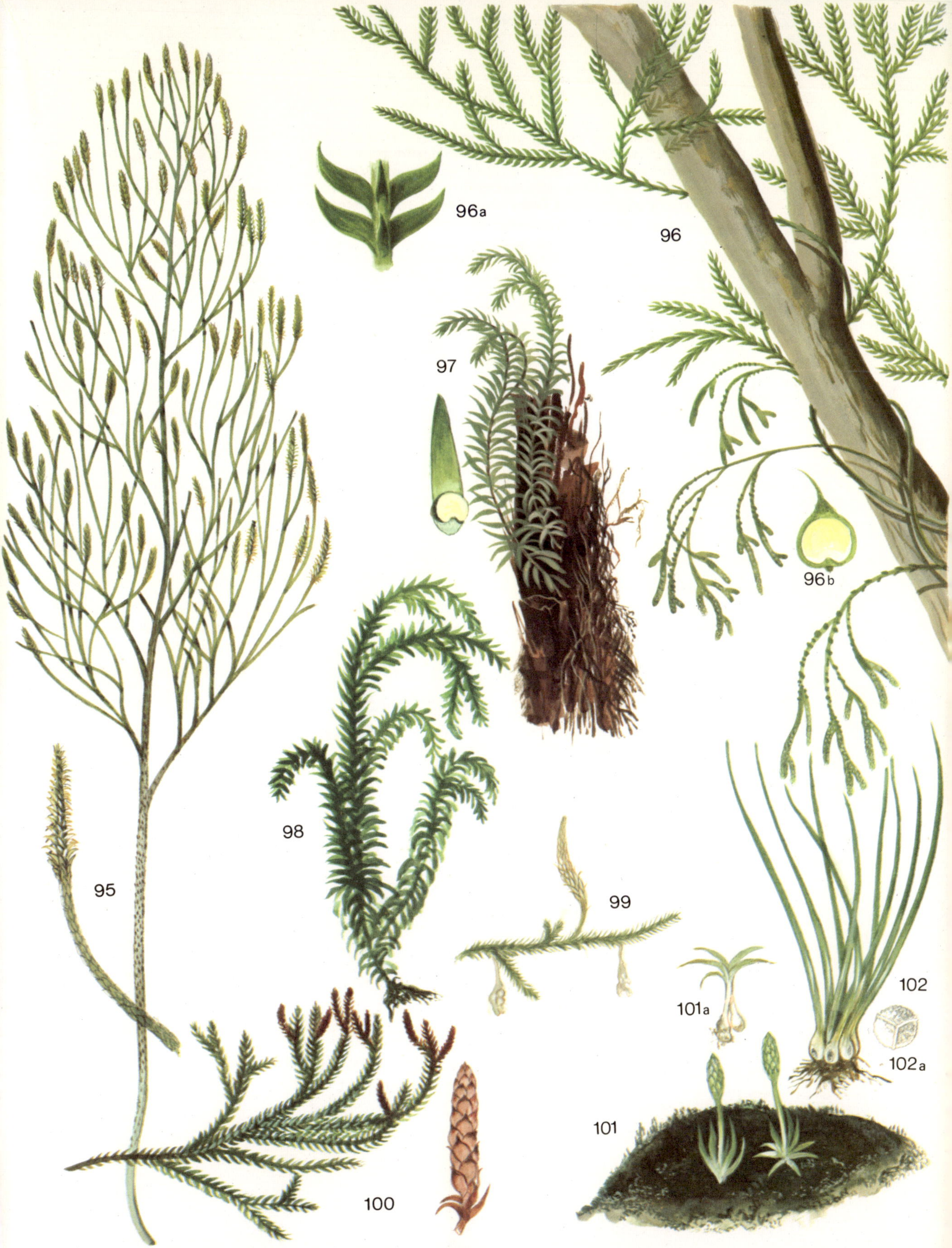
96a
96
97
96b
98
95
99
102
101a
102a
101
100

PSILOTOPSIDA

THIS SMALL GROUP has always been of great interest to botanists because the two genera that comprise it, *Psilotum* and *Tmesipteris*, are the simplest constructed vascular plants living today. They do not possess roots but merely a subterranean rhizome and an aerial stem system which are structurally very similar. The aerial stems have spirally arranged appendages ("leaves"), which are scale-like in *Psilotum* and leafy in *Tmesipteris*, and fused sporangia at the ends of short lateral branches. The structure that the fused sporangia form is called the synangium and this is composed of two sporangia in *Tmesipteris* and three in *Psilotum.*

There are two species of *Psilotum* which are widespread throughout the tropics and subtropics and about twelve species of *Tmesipteris* which are more restricted in their distribution and are largely confined to Australasia and New Caledonia although a few species extend northwards to New Guinea and the Philippines and eastwards to French Polynesia.

103. Psilotum nudum. Aerial stems 5-50 cm long, dichotomously branched many times in the upper part. Appendages ("leaves") small, scale-like, spirally arranged. Synangium trilocular, terminating short lateral branches, green at first, yellow at maturity, brown when old. Two leaves are associated with the synangium (103a). (**103.** x0.7; **103a.** x15)

Restricted to the North Island where it is found in scattered places from Lake Taupo northwards. It is particularly common in some thermal areas such as Waimangu Thermal Reserve where it thrives on the warm soils and on cliffs.

104 Tmesipteris elongata. Aerial stems simple or dichotomously branched one to three times, 10-80 cm or more long. Leafy appendages 1-4 cm long, dull green, flexible. Synangium elongate, rounded at the ends, dropping off when old. (**104.** x0.8; **104a.** x4)

Common throughout the North Island and in the north and west of the South Island, Stewart Island and the Chatham Islands. Usually found on tree-fern trunks.

105. Tmesipteris tannensis. Aerial stems simple, 10 to 80 cm long. Leafy appendages 0.5-3 cm long, clustered together, deep green with a shiny upper surface, stiff, brittle. Synangium tapering at both ends, sporangia at an angle to each other. (**105.** x4. Fertile branch only)

Abundant throughout the country as an epiphyte on trees or on organic accumulations on banks or at the base of trees. In the North Island it is more common at altitudes above 200 metres.

106. Tmesipteris lanceolata. Aerial stems simple, 5-20 cm long. Fertile branches spirally arranged, confined to the lower half of the stem. Sterile leafy appendages flattened into one row, ovate with an acute tip, the upper surface deep green, glossy, lower surface dull green. Synangium small rounded. Aerial stems terminated by a sterile appendage. (**106.** x1)

This small species has been overlooked in the past. It is quite common in the northern North Island and extends as far south as Taranaki where it is known to occur at Tarata and at Tataraimaka south of New Plymouth. Usually found growing on tree-fern trunks but occasionally in the thick mat of aerial rootlets at the base of *Cyathea medullaris* (Black Tree-fern).

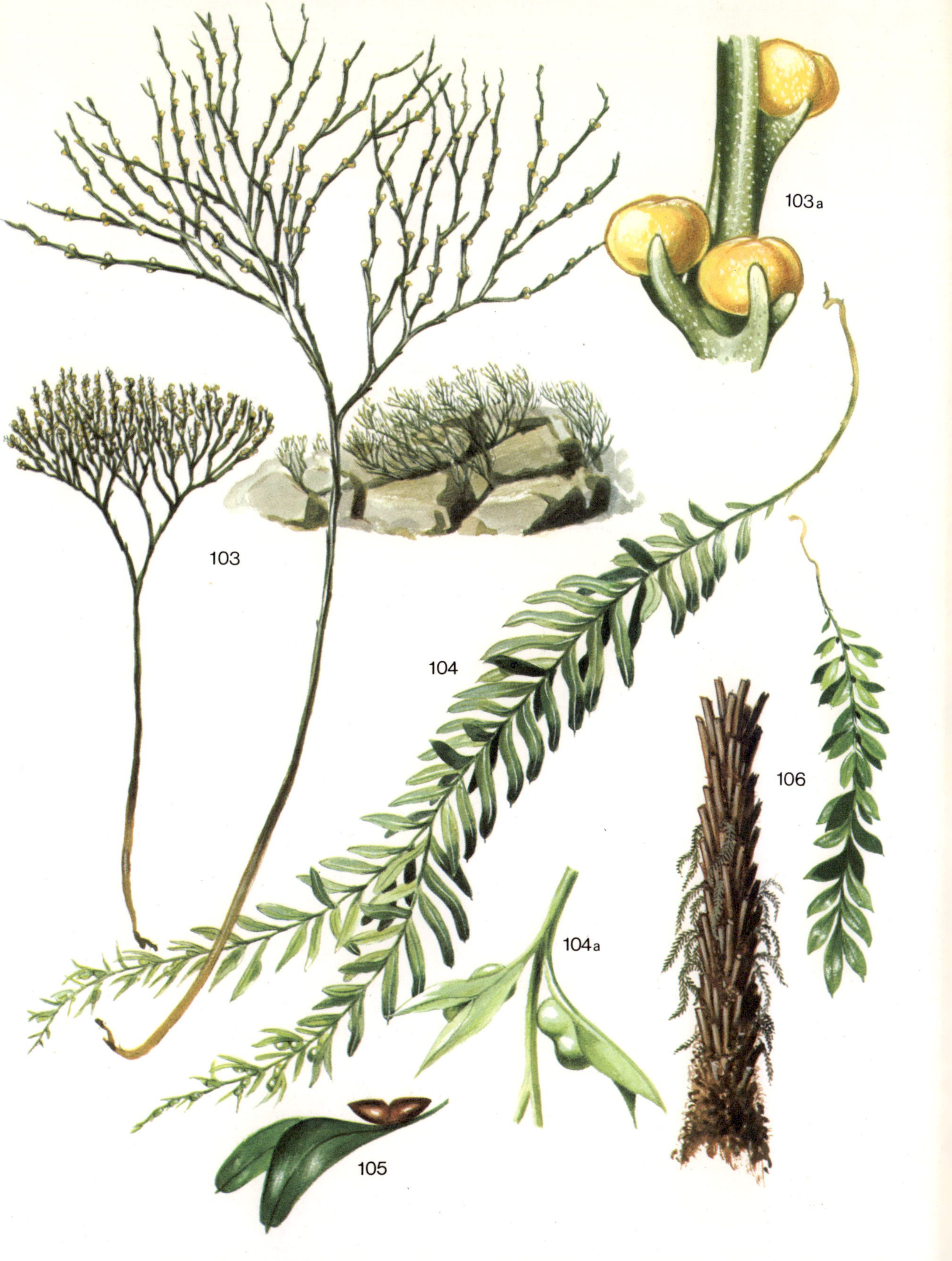

103
103a
104
104a
105
106

CULTIVATION AND PROPAGATION OF FERNS AND FERN ALLIES

Cultivation

It would be an impossible task for me to try and cover completely in a few paragraphs the cultivation and propagation of the New Zealand ferns for these topics alone are sufficient to fill a book. Each species has its own preferences, soil, light, moisture and temperature tolerances and when brought into cultivation it is essential to try and recreate as far as possible the natural conditions under which the species grows. A species which is found in a broad range of habitats will be much easier to cultivate than a species with a very specific one. Thus, *Asplenium bulbiferum* and *Adiantum hispidulum* are very easy to grow while *Tmesipteris* and *Psilotum* are much more difficult. A sun-loving fern such as *Pteridium esculentum* (bracken) thrives in open situations and on the margin of forest but it readily succumbs under very shaded conditions. Fortunately, many of our ferns are very hardy and most of the Spleenworts, Shield-ferns, Tree-ferns, Maidenhair-ferns and Blechnum ferns can be grown successfully without too much difficulty.

Many of our ferns are suitable for indoor culture and it is preferable to select species with tufted or slow creeping rhizomes otherwise there is often difficulty retaining them. Each gardening book or Gardener will give you their "special fern mix" but I have found that a simple mixture of sand, leaf-mould (peat or compost) and loam in equal proportions will satisfy most ferns. One can add gravel or broken brick to the mixture as this helps to maintain good drainage. It is essential to keep pot-grown ferns moist all the time and not allow them to dry out. Erratic watering will lead to disaster. Plastic containers are preferable to clay ones as water is not lost through the walls of the pot.

The garden provides much more scope, and the presence of a shade house or a sheltered corner with a good tree cover will enable you to grow most ferns outside. If the soil is heavy, add coarse sand to break it up. Most ferns prefer organic matter in their soil and as most gardens lack an organic ground cover, unlike natural soils under forest, addition of leaf-mould (peat or compost) is worthwhile.

Most ferns prefer shade, so the south facing aspects can be utilised especially if you do not possess a shade house or shelter belt under trees. Many ferns in the open will be affected by frosts so care must be taken to protect from this. Tree-ferns, especially *Cyathea dealbata* and *Dicksonia squarrosa*, can be grown in the open garden as both will tolerate full sun as long as they receive good water supplies. Ferns that are typically found along streams and on wet banks will require very damp situations and unless rainfall is high you will have to compensate in some way by creating a small pond or by frequent waterings. Evidence of successful cultivation is in seeing the fern fronds increase in size.

The rockery is suitable for a number of our ferns including *Cheilanthes sieberi* and *Nephrolepis cordifolia* as these occur naturally in rocky places and on banks. These species will thrive quite happily in sunny situations as their roots will keep moist and cool amongst the rocks. In sheltered rockeries with plenty of shade a far greater number of species can be grown.

The shade house or a sheltered position under trees is ideal for establishing a fernery. Tree-ferns should be planted if there is sufficient room as these will provide added protection for more delicate ferns and their trunks, when developed, can be utilised by growing epiphytic species on them. Once the fernery has become established with tree-ferns and ground ferns, filmy ferns can be tried. The filmy ferns together with the Crape fern are some of the most difficult ferns to grow. Their delicate, very thin, fronds readily dry out and to establish them successfully a high humidity must be attained. Some species of filmy fern are much more difficult than others. *Hymenophyllum demissum, H. scabrum, H. sanguinolentum* and *Trichomanes reniforme* are not too difficult and these species should be attempted first. Other species such as *Trichomanes endlicherianum, T. elongatum* and *Hymenophyllum ferrugineum* are very difficult and success will only be attained if a plastic or glass frame is built to increase the humidity. Another alternative is to have a small pool in the fernery with steep banks around it and under these conditions these more difficult filmy ferns and Crape fern should establish.

Finally I wish to emphasise again that a knowledge of a species' capabilities before you cultivate it contributes greatly to success. Before planting ferns out in the garden it is advisable to harden the plants off gradually. A fern which has been first established in a pot, hardened off gradually, and then planted out in the garden, is more likely to flourish than a plant which has been collected and planted direct in the garden. If plants are collected in the wild it is best to collect very young plants on banks as these transplant readily. To collect large plants is not satisfactory as the plants usually die, but on banks, the best source of supply, many of the small fern plants will never grow to maturity because of the competition so the removal of a few will give them a new lease on life. **Never collect in National Parks or Reserves and only collect a species when it is common in the area.**

Ferns for the rockery: *Adiantum aethiopicum, Adiantum hispidulum, Asplenium flabellifolium, Cheilanthes distans, Cheilanthes sieberi, Lycopodium cernuum, Lycopodium volubile, Microsorium diversifolium, Nephrolepis cordifolia, Pellaea rotundifolia, Polystichum richardii, Pyrrosia serpens.*

Ferns for shaded rockeries, south-facing aspects and for ferneries: *Adiantum* species, *Asplenium* species, *Athyrium japonicum, Blechum* species, *Cyathea* species, *Cyclosorus penniger. Dicksonia* species, *Gleichenia* species, *Hypolepsis* species, *Lastreopsis* species, *Microsorium* species, *Pellaea rotundifolia, Polystichum* species, *Pteris* species.

Ferns ideal for indoor pot culture: *Adiantum aethiopicum, Adiantum hispidulum. Asplenium bulbiferum, Asplenium lucidum, Doodia media, Lygodium articulatum* (climber), *Nephrolepis cordifolia, Pteris macilenta, Pteris tremula.*

Propagation

Ferns with long creeping rhizomes like *Microsorium diversifolium* and *Adiantum aethiopicum* may be increased in number by making cuttings of the rhizome. The rhizome of *Microsorium* should be cut into pieces containing a short lateral shoot with a frond. *Adiantum aethiopicum* has a thin wiry rhizome which produces tufts of fronds at intervals and in this case sections of the rhizome that possess tufts of fronds can be separated and potted up. Cuttings should be placed in a light sandy soil in a sheltered cool position.

Asexual bulbils like those found in *Asplenium bulbiferum* will not develop in nature until the frond has rotted or if the frond is prostrate in the ground. If a portion of a frond containing bulbils is placed in a pot and lightly covered with soil the bulbils will produce roots in a few weeks and start to develop. They can then be separated and potted up individually.

Growing ferns from spores is by far the most interesting and satisfying method of obtaining plants. It is advisable for the beginner to start with easy species such as *Pteris tremula, Histiopteris incisa* and *Cyathea dealbata* and once you have become familiar with the techniques more difficult species can be attempted.

To obtain spores, collect a portion of a mature frond and place it on a piece of paper in a dry place with the sori face down. If it is at the right stage, the paper will be covered with a brown, black or yellow powder which is a mixture of spores and sporangial fragments.

There are numerous methods of growing ferns from spores and two are briefly outlined here. Fill a small clay pot with sphagnum moss (obtainable from garden centres) and sterilise with boiling water. It is essential to sterilise materials including soil to destroy unwanted organisms such as blue-green algae, fungi and insects as these can wipe out a gametophyte colony. Invert the pot and place in a bowl of water and when the pot surface is moist sprinkle a small amount of spore powder over the pot. Seal the pot by covering with a jar (see diagram). After a few weeks gametophytes will become visible on the surface. Water sterilised by boiling or by adding a small amount of potassium permanganate should be added to the bowl as required and the jar should not be removed as airborne spores may be introduced. When sporophytes become visible prick off the pot when they have produced their second frond.

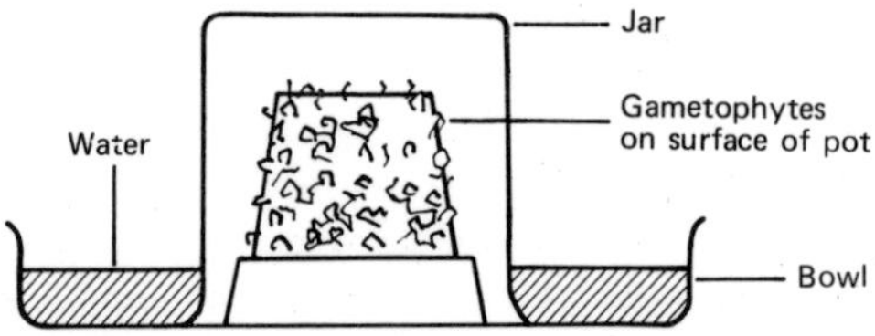

An alternative method is to place sterilised soil in a pot leaving a gap of four to five centimetres below the rim. Sprinkle the spore material over the surface and cover the pot with a piece of glass or seal with plastic. Gametophytes will appear from about three weeks onwards depending on the species. In both methods it is advisable to "black out" the pots for about ten days. Cover with four sheets of paper and place in a shaded situation. After ten days gradually remove the paper one sheet at a time over a few days. Keep in a shaded place. A good account of both cultivation methods and propagation of ferns is found in Reginald Kaye's *Hardy Ferns* and although this book is on ferns cultivated in Britain, many New Zealand species are mentioned, and the methods of cultivation can be applied here in New Zealand.

GLOSSARY

alternate: placed singly along an axis, not in opposite pairs
appressed: lying flat against the organ
aquatic: living in water
asexual: non-sexual
bulbil: a young plant which develops asexually on the leaf or axis of a plant
dichotomous: branching by forking into pairs
dimorphic: occurring in two forms
epidermis: the "skin" or outermost layer of tissue
epiphyte: a plant that grows on another plant but not parasitically
flexuose: zig-zag, wavy
frond: the fern leaf
glabrous: smooth, lacking hairs or scales
glandular: furnished with glands—said of hairs with a swollen terminal head
habitat: the place in which the plant grows
hair: a thin slender outgrowth of the epidermis consisting of a single row of cells
indusium: outgrowth of the epidermis covering the sorus
lanceolate: lance-shaped
lamina: the blade of the frond
marginal: on, or close to the margin
membranous: thin, more or less flexible and translucent
microphyll: the sterile Lycopod leaf
oblique: slanting
panicle: an indeterminate branched structure with numerous strobili
pinnatifid: lobed in a pinnate manner
pinnule: ultimate stalked division of a pinna
pteridophyte: a general name for the ferns and fern-allies
rachis: the midrib of the fern frond (pl. raches)
scale: a thin flat outgrowth of the epidermis, a number of cells wide
segment: an individual free part of an organ
serrate: with sharp teeth
sorus: a group of sporangia
spike: the fertile pinnae of Ophioglossum and Botrychium
sporangium: a sac or other structure containing spores
sporophyll: the specialised Lycopod leaf bearing a sporangium at, or near, its base
synangium: a structure which consists of two or more fused sporangia
stellate: star-shaped
stipe: the stalk of the fern frond
stolon: a stem which runs along the ground and forms a new plant at its tip
strobilus: a cone, in the Lycopods the specialised structure upon which the sporophylls are grouped together
subterranean: below the ground
superficial: occurring on the surface away from the margin
taxon (pl. taxa): a taxonomic group of any rank such as a genus, species, subspecies variety, etc. etc.
translucent: almost transparent
undulate: wavy in a plane at right angles to the surface
zygote: the fertilised egg before it undergoes differentiation

SUGGESTED BOOKS FOR FURTHER READING

ALLAN, H.H. *The Flora of New Zealand*, Vol. 1, Government Printer, Wellington, 1961.
CROOKES, M.A. and DOBBIE, H.B. *New Zealand Ferns*, Whitcombe & Tombs Ltd, 1963.
KAYE, R. *Hardy Ferns*, Faber & Faber Ltd., London, 1968.
SPORNE, K.R. *The Morphology of Pteridophytes*, Hutchinson & Co. Ltd;. London, 1966.
STEVENSON, G. *A Book of Ferns*, Pauls Book Arcade, Hamilton, N.Z., 1959.

INDEX